OXFORD **IB STUDY GUIDES**

Andrew Allott

Biology

FOR THE IB DIPLOMA

OXFORD
UNIVERSITY PRESS

OXFORD
UNIVERSITY PRESS

Great Clarendon Street, Oxford OX2 6DP

Oxford University Press is a department of the University of Oxford.
It furthers the University's objective of excellence in research,
scholarship, and education by publishing worldwide in

Oxford   New York

Auckland Cape Town Dar es Salaam Hong Kong Karachi
Kuala Lumpur Madrid Melbourne Mexico City Nairobi
New Delhi Shanghai Taipei Toronto

With offices in

Argentina Austria Brazil Chile Czech Republic France Greece
Guatemala Hungary Italy Japan Poland Portugal Singapore
South Korea Switzerland Thailand Turkey Ukraine Vietnam

Oxford is a registered trade mark of Oxford University Press
in the UK and in certain other countries

British Library Cataloguing in Publication Data

Data available

ISBN: 978-0-19-838994-1
10 9 8 7 6 5 4 3 2

Printed in Great Britain by Bell & Bain Ltd, Glasgow

Paper used in the production of this book is a natural, recyclable product made from wood grown in
sustainable forests. The manufacturing process conforms to the environmental regulations of the country
of origin

Acknowledgments

We are grateful to the following to reproduce the following copyright material.

pp5, 38, 155 Andrew Allott; p24 CNRI/Photolake Inc; p25 Dennis Kunkel/ Photolake Inc; p37 reproduced
with permission from *The Complete Work of Charles Darwin Online* http://darwin-online.org.uk; p45 (top and
middle) Ben Osborne; p91 James Mendelssohn; p94 John and Lewis, *The Meiotic Mechanism*, Oxford Biology
Reader No 65 (OUP); p95 John and Lewis, *The Meiotic Mechanism*, Oxford Biology Reader No 65; p99 Quest/
Science Photo Library; p103 Microscopix; p120 Science Photo Library/Susan Leavines; p138 Wellcome/
Science Photo Library/Photolibrary; p141 Dr DN Furness; p157 (left and right) Alison Allott; p163 Steve
Gschmeissner/ Science Photo Library/Photolibrary; p165 Innerspace Imaging/Science Photo Library.

Cover photo: Natural Visions/Alamy

We have tried to trace and contact all copyright holders before publication. If notified the publishers will be
pleased to rectify any errors or omissions at the earliest opportunity.

MIX
Paper from
responsible sources
FSC® C007785

Introduction and acknowledgements

The IB Biology Programme continues to evolve, in response to developments in research and changes in the perceived importance of different topics. This book is intended to help students find the information that they need quickly and easily and allow time for thinking about the issues involved. It follows the programme being taught from 2007 that will be examined from 2009 onwards.

All topics needed for Higher Level (HL) and Standard Level (SL) Biology are covered, including all eight options. The topics covered are in the same sequence as in the syllabus, but within some topics the sequence of sub-topics has been slightly altered, to give a more coherent progression of ideas. An index of assessment statements in the syllabus has been included to show where each one is covered.

- Topics 1–6 are core topics studied at both HL and SL.
- Topics 7–11 are additional topics studied only at HL.
- Options A–C are studied only at SL (each student takes two of the eight options).
- Options D–G can be studied at HL or SL, with extra material needed at HL, separated on clearly marked pages at the end of the option.
- Option H is studied only at HL.

Practice questions are included at the end of topics. Most of these are taken, with permission, from past IB examination papers. Answers to each question are given, though students and teachers may be able to find other valid answers!

Guidance is given for students working on internal assessment, extended essays or preparing for final exams.

There has never been a more important time to study biology. There are unprecedented opportunities for using recently developed techniques in beneficial ways, but there are also greater threats to the natural world than for millions of years. A thorough understanding of the principles of biology is essential if we are to counter the threats and make the most of the opportunities. Biology teachers worldwide should continue to be commended for the work they do in promoting this understanding. Teachers of IB Biology often take on an additional challenge – to promote international understanding. There are many opportunities for this in Biology. Apart from humans, living organisms do not recognize national frontiers. Living organisms throughout the biosphere, including humans, are interdependent. Human activities have international impacts, so international co-operation is essential to protect the biosphere and its treasure-house of biodiversity.

I am very grateful for the help that fellow teachers have given me during the writing of this book. Heather Addison was also a great help in her work as Editor. I am indebted to my wife Alison and son William for their support and forbearance during the many hours that I have spent on it. I would like to dedicate the work that I did on the book to all biologists around the world, who are striving to conserve living organisms and their habitats.

CONTENTS

(Italics denote Higher level pages)

ORIGIN OF INDIVIDUAL QUESTIONS

The questions detailed below are all taken from past IB examination papers. The questions are from May (M) or November (N), Sample, 1998 (98), 1999 (99) and 2000 (00) paper 2 (P2) or paper 3 (P3) with question number in brackets. All other questions are IB style questions written by the author of this book.

TOPICS 1 AND 2 STATISTICAL ANALYSIS AND CELLS
1. MOOSLP2(1)   **2.** M99SLP2(1)

TOPIC 3 THE CHEMISTRY OF LIFE
3. N99SLP2(3)

TOPIC 4 GENETICS
1. SAMPLE HLP2(3)   **2.** M00SLP2(2)

TOPIC 5 ECOLOGY AND EVOLUTION
1. M99SLP2(1)   **2.** M98HLP2(3)

TOPIC 6 HUMAN HEALTH AND PHYSIOLOGY
1. N99SLP2(1)   **2.** N99SLP2(2)

TOPIC 7 NUCLEIC ACIDS AND PROTEINS
3. N98HLP2(3)

TOPIC 8 CELL RESPIRATION AND PHOTOSYNTHESIS
1. N98SLP3(C1)   **2.** N99SLP3(C2)

TOPIC 9 PLANT SCIENCE
3. M98SLP3(C1)

TOPIC 10 GENETICS
1. M99HLP2(3)   **2.** N98HLP2(2)

TOPIC 11 HUMAN HEALTH AND PHYSIOLOGY
Questions written by author

OPTION A – HUMAN NUTRITION AND HEALTH
Questions written by author

OPTION B – PHYSIOLOGY OF EXERCISE
1. N98SLP3(B1)

OPTION C – CELLS AND ENERGY
1. SAMPLE SLP3(C1)   **2.** M98SLP3(C2)

OPTION D – EVOLUTION
1. M98HLP3(D1)   **3.** N98HLP3(D2)

OPTION E – NEUROBIOLOGY AND BEHAVIOUR
1. M99HLP3(E2)   **2.** N99SLP3(E2)

OPTION F – MICROBES AND BIOTECHNOLOGY
Questions written by author

OPTION G – ECOLOGY AND CONSERVATION
1. N00SLP3(G1)   **2.** M98SLP3(G3)

OPTION H – FURTHER HUMAN PHYSIOLOGY
2. N99HLP3(H2)

Mean and standard deviation

There is almost always variation in biological data. The amount of variation can be shown using a graph called a frequency distribution. Most variation gives a bell-shaped frequency distribution called the **normal distribution**. The mean value is in the middle of the distribution. The **mean** of a set of values is calculated by dividing the sum of the values by the number of values.

For example, the sum of the values 7, 9, 11 and 17 is 44 and as there are four values, the mean is 44 divided by 4, which is 11.

The **standard deviation** is used to assess how far the values are spread above and below the mean. It is calculated by entering data into a graphic display or scientific calculator and pressing the standard deviation function key. A high standard deviation shows that the data are widely spread, whereas a low standard deviation shows that the data are clustered closely around the mean.

The standard deviation can be used to help decide whether the difference between two means is likely to be significant. Two examples are described below.

The normal distribution

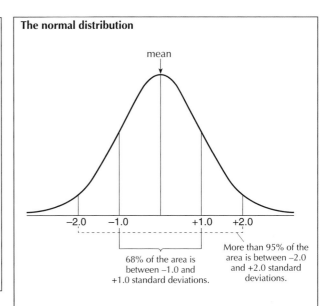

A useful rule is that 68% of the values lie within one standard deviation of the mean in a normal distribution and approximately 95% of the values lie within two standard deviations of the mean (above).

ERROR BARS

Bars on graphs extending above and below the mean value are used to show the variability of the data. They may show the range of the data, or the standard deviation.

LEFT AND RIGHT HAND LENGTHS

Thirty teenage boys measured the length of their left and right hands, to find out whether they are different. Individual boys' left and right hand length varied by as much as 10mm. The results are shown in the frequency distribution below.

Hand	Mean length	Standard deviation
left	188.6 mm	11.0 mm
right	188.4 mm	10.9 mm

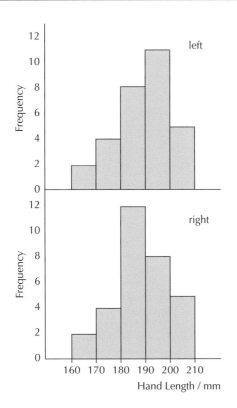

Because the standard deviations are much greater than the difference in mean length, it is very unlikely that the difference in mean length between left and right hands is significant.

HAND AND FOOT LENGTHS

The same thirty teenage boys who measured their hand lengths also measured the length of their right foot, to find out whether it was different from their hand lengths. The results are shown in the frequency distribution below.

Hand/foot	Mean length	Standard deviation
right hand	188.4 mm	10.9 mm
right foot	262.5 mm	14.3 mm

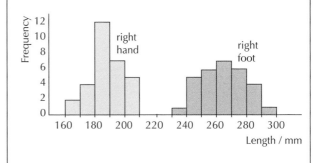

Because the standard deviations are much less than the difference in mean length, it is very likely that the difference in mean length between right hands and right feet is significant.

Relationships – significance and cause

THE *t*-TEST

On the previous page, sizes of standard deviations were used to assess whether differences between means were likely to be significant. Biologists often need to decide more objectively whether differences between means are significant. One of the most frequently used methods is called the *t*-test.

The *t*-test can be used to find out whether there is a significant difference between the means of two populations. A difference is considered statistically significant if the probability of it being due to random variation is 5% or less.

t is a statistic that is calculated from the two sets of measurements. The larger the difference between the two means, the larger *t* is. The larger the standard deviations, the smaller *t* is. It is not necessary to learn how to calculate *t*, because a graphic display calculator or computer is nearly always now used.

Stages in using the *t*-test

1. Enter the values in a graphic display calculator or a spreadsheet program, with values for the two populations entered separately.
2. Use the calculator function keys or computer software to calculate *t*.
3. Find the number of degrees of freedom. This will be the total number of values in both populations, minus two.
4. Find the critical value for *t* either using computer software or a table of values of *t*. The level of significance (*P*) chosen should be 0.05 (5%) and the appropriate row should be selected according to the number of degrees of freedom.
5. Compare the calculated value of *t* with the critical value. If the critical value is exceeded, there is evidence of a significant difference between the means, at the 5% level.

Table of critical values of *t*

	\multicolumn{6}{c}{Level of significance (*P*)}
Degrees of freedom	0.2	0.1	0.05	0.02	0.01	0.002
1	3.078	6.314	12.706	31.821	83.657	318.310
2	1.886	2.920	4.303	6.985	9.925	27.327
3	1.638	2.353	3.182	4.541	5.841	10.215
4	1.533	2.132	2.776	3.747	4.604	7.173
5	1.476	2.015	2.571	3.365	4.032	5.893
6	1.440	1.943	2.447	3.143	3.707	5.208
7	1.415	1.895	2.385	2.998	3.499	4.785
8	1.397	1.860	2.308	2.896	3.355	4.501
9	1.383	1.833	2.262	2.821	3.250	4.297
10	1.372	1.812	2.228	2.764	3.169	4.144
11	1.363	1.796	2.201	2.718	3.106	4.025
12	1.356	1.782	2.179	2.681	3.055	3.930
13	1.350	1.771	2.180	2.650	3.012	3.852
14	1.345	1.761	2.145	2.624	2.977	3.787
15	1.341	1.753	2.131	2.602	2.947	3.733
16	1.337	1.746	2.120	2.583	2.921	3.686
17	1.333	1.740	2.110	2.567	2.898	3.646
18	1.330	1.734	2.101	2.552	2.878	3.610
19	1.328	1.729	2.093	2.539	2.861	3.579
20	1.325	1.725	2.086	2.528	2.845	3.552
30	1.310	1.697	2.042	2.457	2.750	3.385
40	1.303	1.684	2.021	2.423	2.704	3.307

Examples of the use of the *t*-test

These examples are based on the data for hand and foot lengths described on the previous page.

1. Testing the difference between mean lengths of left and right hands
   - Mean length of left hands = 188.6mm
   - Mean length of right hands = 188.4mm
   - *t* = 0.082
   - Critical value for *t* = 2.002 (*P* = 0.05)

The calculated value of *t* is much smaller than the critical value, so the difference between the mean lengths of left and right hands is not significant.

2. Testing the difference between the mean lengths of right hands and right feet.
   - Mean length of right hands = 188.4mm
   - Mean length of right feet = 262.5mm
   - *t* = 23.3
   - Critical value for *t* = 2.005 (*P* = 0.05)

The calculated value of *t* is much larger than the critical value, showing that the difference between the mean lengths of hands and feet is significant.

In these two examples, the *t*-test confirms conclusions that are reasonably obvious. In biological research, it is often much less clear whether differences between means are significant and the *t*-test is therefore very useful.

CORRELATION AND CAUSE

The scattergraph below shows that there is a positive correlation between the lengths of the right hand and right feet of thirty teenage boys – boys with larger hands tend to have larger feet as well.

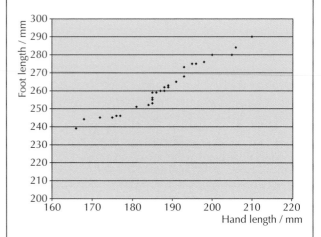

Although there is a positive correlation between hand and foot length, we know that increases in the length of the hand do not cause increases in length of the foot. Instead, both are due to the factors that control growth in teenage boys. This mistake is often made in analysis of data – a correlation between two variables is assumed to show that there is a causal link. It is important to remember that correlation is not proof of cause.

Cell theory

INTRODUCING CELLS

Cells consist of **cytoplasm**, enclosed in a **plasma membrane**, usually controlled by a single **nucleus**. Two cell types that can be easily looked at under a light microscope are human cheek cells, scraped from inside the mouth (left) and moss leaf cells (right).

Human cheek cell

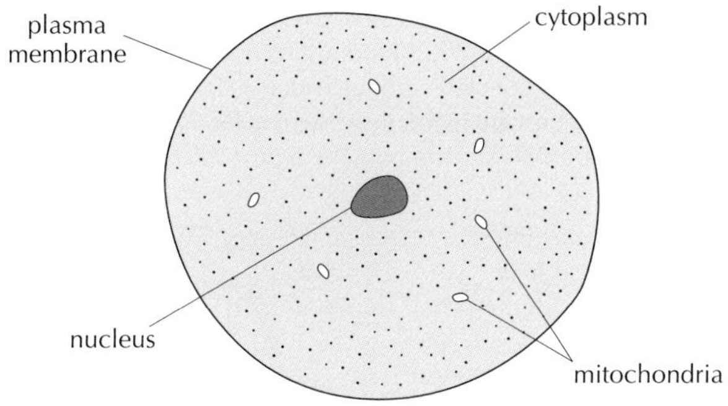

plasma membrane
cytoplasm
nucleus
mitochondria

Moss leaf cell

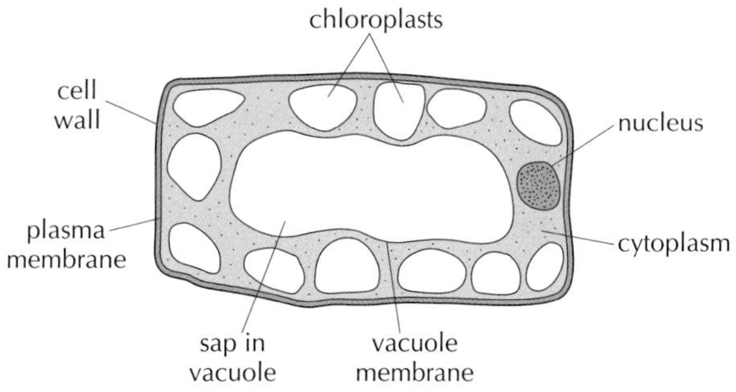

chloroplasts
cell wall
plasma membrane
nucleus
cytoplasm
sap in vacuole
vacuole membrane

UNICELLULAR ORGANISMS

Some organisms such as *Amoeba* (below), *Chlorella* and *Euglena* have only one cell. This single cell has to carry out all the functions of life:

metabolism – chemical reactions inside the cell
response – reacting to stimuli
homeostasis – controlling conditions inside the cell
growth – increasing in size
reproduction – producing offspring
nutrition – obtaining food.

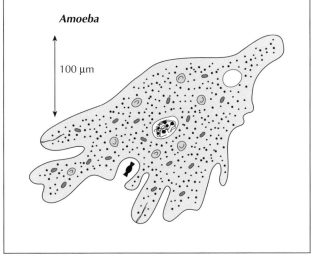

Amoeba

100 μm

MULTICELLULAR ORGANISMS

Multicellular organisms consist of many cells. These cells do not have to carry out many different functions. Instead, they can become specialized for one particular function and carry it out very efficiently. Cells in a multicellular organism therefore develop in different ways. This is called **differentiation**. The way in which this occurs is described on page 4.
Multicellular organisms are said to show **emergent properties**. This means that the whole organism is more than the sum of its parts, because of the complex interactions between cells.

THE CELL THEORY

The cell theory includes these statements:
• living organisms are composed of cells
• cells are the smallest units of life
• cells come from pre-existing cells.

Many organisms have been examined and have been found to consist of cells, but there are some cases where the idea of living organisms consisting of tiny box-like structures does not seem to fit. For example, skeletal muscle is made up of **muscle fibres**. These are much larger than most cells (300 or more mm long) and contain hundreds of nuclei. Most fungi consist of thread-like structures called **hyphae**, which in some species contain many nuclei without dividing walls between. Many tissues, such as bone, contain a greater volume of **extracellular material** (material outside the cell membrane) than of cells. Despite these awkward cases, most living tissues are composed of cells. Also, whereas cells taken from an organism often survive for a time, smaller parts of an organism do not. Cells do therefore seem to be the smallest units of life that are capable of survival.

There is also evidence for the third part of the cell theory. Some of the classic experiments in biology showed that spontaneous generation of life is impossible (below). The first cells must have been formed in the origin of life from non-cellular material, but today there is no evidence that cells can be formed except by cell division.

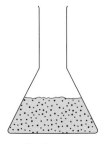

Sterilized soup in an open container decays because bacteria float in

Sterilized soup in a sealed container does not decay as no bacteria are present

Stem cells and differentiation

DIFFERENTIATION

Cells in a multicellular organism develop in different ways and can therefore carry out different functions. This is called **differentiation**. The cells need different genes to develop in different ways. Each cell has all of these genes, so could develop in any way, but it only uses the ones that it needs to follow its pathway of development. Once a pathway of development has begun in a cell, it is usually fixed and the cell cannot change to follow a different pathway. The cell is said to be committed. The drawings (below) show three of the hundreds of different types of differentiated cells in humans.

Heart muscle tissue

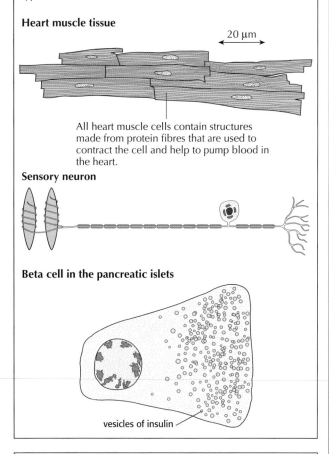

20 µm

All heart muscle cells contain structures made from protein fibres that are used to contract the cell and help to pump blood in the heart.

Sensory neuron

Beta cell in the pancreatic islets

vesicles of insulin

STEM CELLS

Stem cells are defined as cells that have the capacity to self-renew by cell division and to differentiate. Human embryos consist entirely of stem cells in their early stages, but gradually the cells in the embryo commit themselves to a pattern of differentiation. Once committed, a cell may still divide several more times, but all of the cells formed will differentiate in the same way and so they are no longer stem cells.

Small numbers of embryonic cells remain as stem cells however and they are still present in the adult body. They are found in most human tissues, including bone marrow, skin and liver. They give some human tissues considerable powers of regeneration and repair. The stem cells in other tissues only allow limited repair – brain, kidney and heart, for example. There has been great interest in stem cells because of their potential for tissue repair and for treating a variety of degenerative conditions. For example, Parkinson's disease, multiple sclerosis and strokes all involve the loss of neurons or other cells in the nervous system. Although still only at the research stage, there is the potential to use stem cells to replace them.

THERAPEUTIC USE OF STEM CELLS

In the future, many therapies may involve the use of stem cells. Some therapeutic uses have already been introduced. One example is given here.

1. The placenta and umbilical cord of a baby is used as a source of stem cells. At the end of childbirth, the placenta is taken and placed on a stand, with the umbilical cord hanging down from it. Blood drains out of the umbilical cord and is collected – about $100cm^3$. The cord blood contains many hematopoietic stem cells. These cells can divide and differentiate into any type of blood cell.

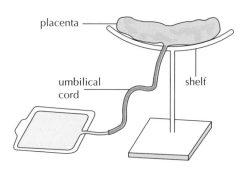

placenta

umbilical cord

shelf

2. Red blood cells are removed from the cord blood and the remaining fluid is then tested to find its tissue type, checked for disease-causing organisms and stored in liquid nitrogen, in a special bank of cord blood.

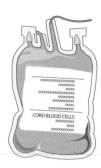

CORD BLOOD CELLS

3. Cord blood can be used to treat patients, especially children, who have developed certain forms of leukemia. This is a cancer in which the cells in bone marrow divide uncontrollably, producing far too many white blood cells. The patient's tissue type is matched with cord blood in the bank. If suitable cord blood is available, the patient is given chemotherapy drugs that kill bone marrow cells, including the cells causing the leukemia.

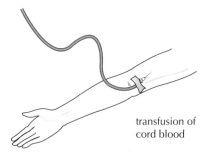

transfusion of cord blood

4. The selected cord blood is taken from the bank, thawed and introduced into the patient's blood system, usually via a vein in the chest or arm. The hematopoietic stem cells establish themselves in the patient's bone marrow, where they divide repeatedly to build up a population of bone marrow cells to replace those killed by the chemotherapy drugs.

Size in cell biology

LIMITATIONS TO CELL SIZE

Cells do not carry on growing indefinitely. They reach a maximum size and then may divide. If a cell became too large, it would develop problems because its surface area to volume ratio would become too small.

As the size of any object is increased, the ratio between the surface area and the volume decreases. Consider the surface area to volume ratio of cubes of varying size as an example. The rate at which materials enter or leave a cell depends on the surface area of the cell. However, the rate at which materials are used or produced depends on the volume. A cell that becomes too large may not be able to take in essential materials or excrete waste substances quickly enough.

The same principle works for heat. Cells that generate heat may not be able to lose it quickly enough if they grow very large.

Surface area to volume ratios are important in biology. They help to explain many phenomena apart from maximum cell sizes.

UNITS FOR SIZE MEASUREMENTS

Most S.I. units differ from each other by a factor of 1000.
One millimetre is a thousand times smaller than 1 metre.
One micrometre is a thousand times smaller than 1 millimetre.
One nanometre is a thousand times smaller than 1 micrometre.
The most useful units for measuring the sizes of cells and structures within them are nanometres (nm) and micrometres (µm).
The typical sizes of some important structures in biology are shown opposite.

CALCULATING MAGNIFICATION

Photographs or drawings of structures seen under the microscope show them larger than they really are – they magnify them. It is useful to know how much larger the image is than the actual specimen. This factor is called the magnification. It is always helpful to show the magnification on a drawing of a biological structure.

Follow these instructions to calculate magnification.
1. Choose an obvious length, for example the maximum diameter of a cell. Measure it on the drawing.
2. Measure the same length on the actual specimen.
3. If the units used for the two measurements are different, convert one of them into the same units as the other one.
4. Divide the length on the drawing by the length on the actual specimen. The result is the magnification.

$$\text{Magnification} = \frac{\text{size of image}}{\text{size of specimen}}$$

This equation can also be used to calculate the actual size of a specimen if the magnification and size of the image are known.

SCALE BARS

A scale bar is a line added to a micrograph or a drawing to help to show the actual size of the structures.

For example, a 10 µm bar shows how large a 10 µm object would appear.

The figure below shows is a scanning electron micrograph of a leaf with the magnification and a scale bar both shown.

Scanning electron micrograph of leaf (× 480)

60 µm

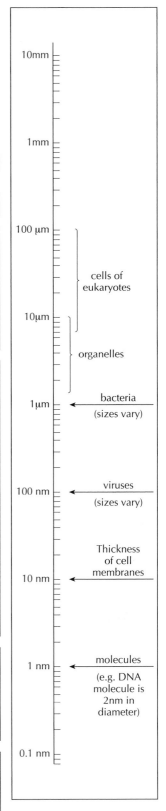

10mm	
1mm	
100 µm	cells of eukaryotes
10µm	organelles
1µm	bacteria (sizes vary)
100 nm	viruses (sizes vary)
10 nm	Thickness of cell membranes
1 nm	molecules (e.g. DNA molecule is 2nm in diameter)
0.1 nm	

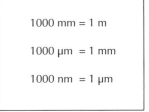

1000 mm = 1 m

1000 µm = 1 mm

1000 nm = 1 µm

Prokaryotic cells

ULTRASTRUCTURE OF CELLS

From the 1950s onwards, cell structure could be studied in much greater detail than before, using electron microscopes. What was revealed is called the ultrastructure of the cell.

Cells were divided into two types according to their structure, **prokaryotic** and **eukaryotic**. The first cells to evolve were prokaryotic and many organisms still have prokaryotic cells, including all bacteria. These cells have no nucleus and the name prokaryotic means *before the nucleus*.

The functions of structures within prokaryotic cells are shown (right). Prokaryotic cells divide in two by a process called **binary fission**.

FUNCTIONS OF PARTS OF A PROKARYOTIC CELL

Structure	Function
Cell wall	Forms a protective outer layer that prevents damage from outside and also bursting if internal pressure is high.
Plasma membrane	Controls entry and exit of substances, pumping some of them in by active transport.
Cytoplasm	Contains enzymes that catalyse the chemical reactions of metabolism and contains DNA in a region called the nucleoid.
Pili	Hair-like structures projecting from the cell wall, that can be ratcheted in and out; when connected to another bacterial cell they can be used to pull cells together.
Flagella	Solid protein structures, with a corkscrew shape, projecting from the cell wall, which rotate and cause locomotion.
Ribosomes	Small granular structures that synthesise proteins by translating messenger RNA. Some proteins stay in the cell and others are secreted.
Nucleoid	Region of the cytoplasm that contains naked DNA, which is the genetic information of the cell.

Electron micrograph of *Escherichia coli* (1–2 µm in length)

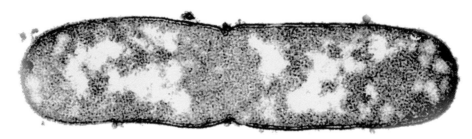

Drawing to help interpret the electron micrograph

ribosomes cell wall plasma membrane cytoplasm nucleoid (region containing naked DNA)

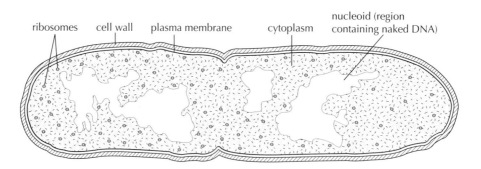

Electron micrograph of *Escherichia coli* showing surface features

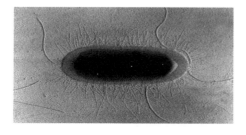

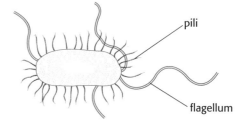

pili

flagellum

Eukaryotic cells

STRUCTURE OF A EUKARYOTIC CELL

Electron micrograph of a liver cell (× 6000)

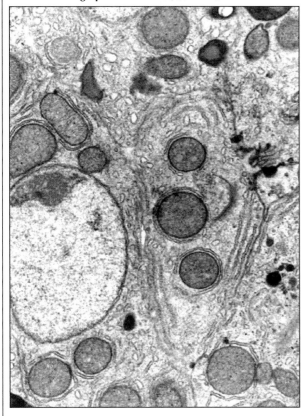

Drawing to interpret parts of the electron micrograph

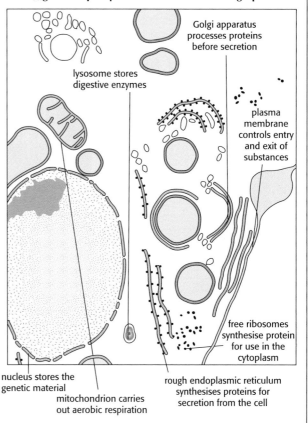

Golgi apparatus processes proteins before secretion

lysosome stores digestive enzymes

plasma membrane controls entry and exit of substances

free ribosomes synthesise protein for use in the cytoplasm

nucleus stores the genetic material

mitochondrion carries out aerobic respiration

rough endoplasmic reticulum synthesises proteins for secretion from the cell

COMPARING PROKARYOTIC AND EUKARYOTIC CELLS

Feature	Prokaryotic cells	Eukaryotic cells
Type of genetic material	A naked loop of DNA	Chromosomes consisting of strands of DNA associated with protein. Four or more chromosomes are present.
Location of genetic material	In the cytoplasm in a region called the nucleoid	In the nucleus inside a double nuclear membrane called the nuclear envelope
Mitochondria	Not present	Always present
Ribosomes	Small sized – 70S	Larger sized – 80S
		(S = Svedberg units – related to the size of organelles)
Internal membranes	Few or none are present	Many internal membranes that compartmentalize the cytoplasm including ER, Golgi apparatuses, lysosomes

COMPARING PLANT AND ANIMAL CELLS

Feature	Animal	Plant
Cell wall	No cell wall, only a plasma membrane	Cell wall *and* plasma membrane present
Chloroplasts	Not present	Present in cells that photosynthesize
Polysaccharides	Glycogen is used as a storage compound	Starch is used as a storage compound
Vacuole	Not usually present	Large fluid-filled vacuole often present
Shape	Able to change shape. Usually rounded	Fixed shape. Usually rather regular

Membrane structure and membrane proteins

Fluid mosaic model of a biological membrane

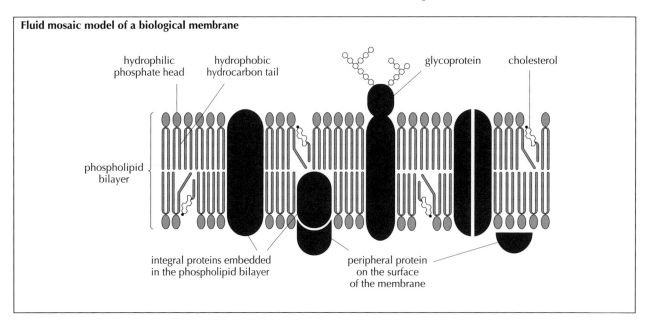

hydrophilic phosphate head

hydrophobic hydrocarbon tail

glycoprotein

cholesterol

phospholipid bilayer

integral proteins embedded in the phospholipid bilayer

peripheral protein on the surface of the membrane

PHOSPHOLIPIDS

Hydrophilic molecules are attracted to water. Hydrophobic molecules are not attracted to water, but are attracted to each other. Phospholipid molecules are unusual because they are partly hydrophilic and partly hydrophobic.

The phosphate head is hydrophilic and the two hydrocarbon tails are hydrophobic. In water, phospholipids form double layers with the hydrophilic heads in contact with water on both sides and the hydrophobic tails away from water in the centre. This arrangement is found in biological membranes. The attraction between the hydrophobic tails in the centre and between the hydrophilic heads and the surrounding water makes membranes very stable.

FLUIDITY OF MEMBRANES

Phospholipids in membranes are in a fluid state. This allows membranes to change shape in a way that would be impossible if they were solid. The fluidity also allows vesicles to be pinched off from membranes or fuse with them.

MEMBRANE PROTEINS

Some electron micrographs show the positions of proteins within membranes. The proteins are seen to be dotted over the membrane. This gives the membrane the appearance of a mosaic. Because the protein molecules float in the fluid phospholipid bilayer, biological membranes are called fluid mosaics. The figure (above) is a diagram showing the fluid mosaic model of a biological membrane. Some of the functions of membrane proteins are shown below.

Functions of membrane proteins

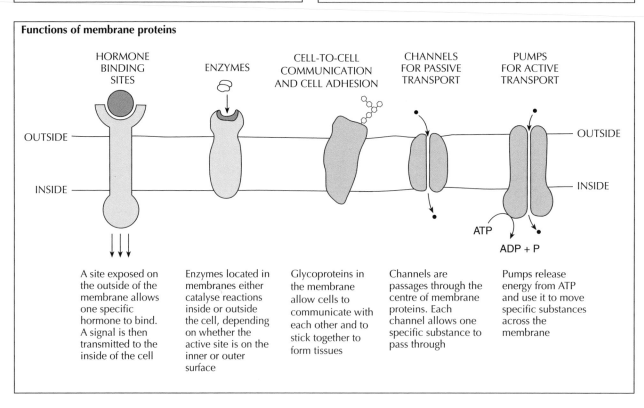

HORMONE BINDING SITES

ENZYMES

CELL-TO-CELL COMMUNICATION AND CELL ADHESION

CHANNELS FOR PASSIVE TRANSPORT

PUMPS FOR ACTIVE TRANSPORT

OUTSIDE

INSIDE

OUTSIDE

INSIDE

ATP

ADP + P

A site exposed on the outside of the membrane allows one specific hormone to bind. A signal is then transmitted to the inside of the cell

Enzymes located in membranes either catalyse reactions inside or outside the cell, depending on whether the active site is on the inner or outer surface

Glycoproteins in the membrane allow cells to communicate with each other and to stick together to form tissues

Channels are passages through the centre of membrane proteins. Each channel allows one specific substance to pass through

Pumps release energy from ATP and use it to move specific substances across the membrane

Passive transport across membranes

DIFFUSION

Solids, liquids and gases consist of particles – atoms, ions and molecules. In liquids and gases, these particles are in continual motion. The direction that they move in is random. If particles are evenly spread then their movement in all directions is even and there is no net movement – they remain evenly spread despite continually moving. Sometimes particles are unevenly spread – there is a higher concentration in one region than another. This causes diffusion.

Diffusion is the passive movement of particles from a region of higher concentration to a region of lower concentration, as a result of the random motion of particles.

Diffusion occurs because more particles move from the region of higher concentration to the region of lower concentration than move in the opposite direction. Diffusion can occur across membranes if there is a concentration gradient and if the membrane is permeable to the particle. For example, membranes are freely permeable to oxygen, so if there is a lower concentration of oxygen inside a cell than outside, it will diffuse into the cell. Membranes are not permeable to cellulose, so it does not diffuse across, even if there is a higher concentration on one side of a membrane than the other.

SIMPLE AND FACILITATED DIFFUSION

Membranes allow some substances to diffuse through but not others – they are **partially permeable**. Some of these substances move between the phospholipid molecules in the membrane – this is **simple diffusion**. Other substances are unable to pass between the phospholipids. To allow these substances to diffuse through membranes, channel proteins are needed. This is called **facilitated diffusion**. Channel proteins are specific – they only allow one type of substance to pass through. For example, chloride channels only allow chloride ions to pass through. Cells can control whether substances pass through their plasma membranes by facilitated diffusion, by the types of channel protein that are produced and inserted into the membrane. Cells cannot control the direction of movement. Facilitated diffusion always causes particles to move from a region of higher concentration to a region of lower concentration. Both simple and facilitated diffusion are passive processes – no energy has to be used by the cell to make them occur.

There are sodium and potassium channel proteins in the membranes of neurones that open and close, depending on the voltage across the membrane. They are called voltage-gated channels and are used during the transmission of nerve impulses.

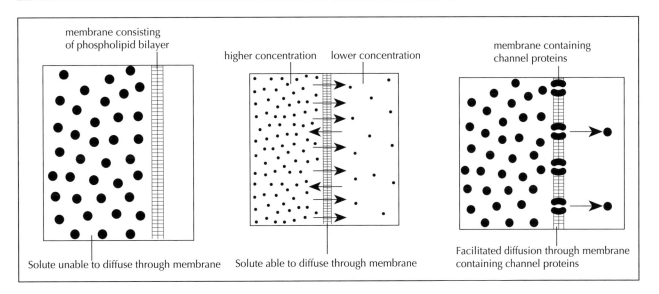

membrane consisting of phospholipid bilayer

Solute unable to diffuse through membrane

higher concentration   lower concentration

Solute able to diffuse through membrane

membrane containing channel proteins

Facilitated diffusion through membrane containing channel proteins

OSMOSIS

Plasma membranes are usually freely permeable to water. The passive movement of water across membranes is different from diffusion across membranes, because water is the solvent. A **solvent** is a liquid in which particles dissolve. Dissolved particles are called **solutes**. The direction in which water moves is due to the concentration of solutes, rather than the concentration of water molecules, so it is called osmosis, rather than diffusion.

Osmosis is the passive movement of water molecules from a region of lower solute concentration to a region of higher solute concentration, across a partially permeable membrane.

Attractions between solute particles and water molecules are the reason for water moving to regions with a higher solute concentration.

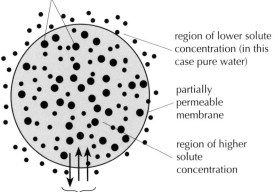

Solute molecules cannot diffuse out as the membrane is impermeable to them

region of lower solute concentration (in this case pure water)

partially permeable membrane

region of higher solute concentration

Water molecules move in and out through the membrane but more move in than out. There is a net movement from the region of lower solute concentration to the region of higher solute concentration

Active transport across membranes

PUMP PROTEINS AND ACTIVE TRANSPORT

Active transport is the movement of substances across membranes using energy from ATP. Active transport can move substances against the concentration gradient – from a region of lower to a region of higher concentration. Protein pumps in the membrane are used for active transport. Each pump only transports particular substances, so cells can control what is absorbed and what is expelled. Pumps work in a specific direction – the substance can only enter the pump on one side and can only exit on the other side.

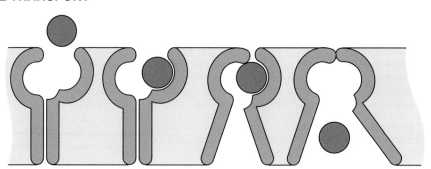

Particle enters the pump from the side with a lower concentration

Particle binds to a specific site. Other types of particle cannot bind

Energy from ATP is used to change the shape of the pump

Particle is released on the side with a higher concentration and the pump then returns to its original shape

TRANSPORT OF MATERIALS BY VESICLES IN THE CYTOPLASM

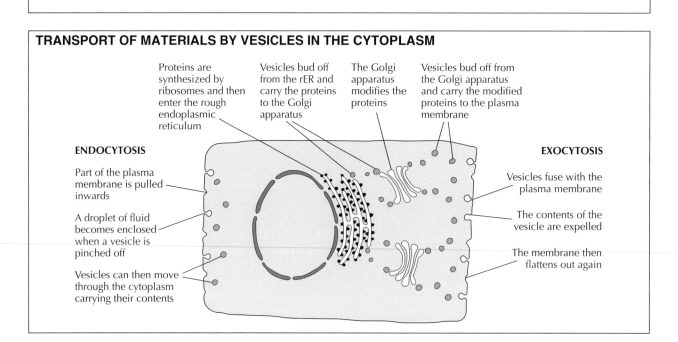

Proteins are synthesized by ribosomes and then enter the rough endoplasmic reticulum

Vesicles bud off from the rER and carry the proteins to the Golgi apparatus

The Golgi apparatus modifies the proteins

Vesicles bud off from the Golgi apparatus and carry the modified proteins to the plasma membrane

ENDOCYTOSIS

Part of the plasma membrane is pulled inwards

A droplet of fluid becomes enclosed when a vesicle is pinched off

Vesicles can then move through the cytoplasm carrying their contents

EXOCYTOSIS

Vesicles fuse with the plasma membrane

The contents of the vesicle are expelled

The membrane then flattens out again

EXTRACELLULAR COMPONENTS

The plasma membrane is the barrier that separates a cell from its surroundings. Cells sometimes produce components and then place them outside the plasma membrane, using exocytosis. These are called extracellular components. Two examples of the roles of extracellular components are outlined here:

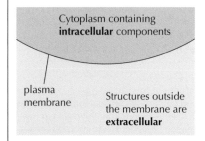

Cytoplasm containing **intracellular** components

plasma membrane

Structures outside the membrane are **extracellular**

1. The plant cell wall

Plants construct their cell walls by synthesising cellulose fibres in vesicles and adding them to the inner surface of the cell wall. Other substances are secreted to interconnect the cellulose fibres. The strength of the cellulose allows plant cell walls to have these roles:
- maintaining the cell's shape
- allowing high pressure to build up in the cell without it bursting
- high pressure in plant cells prevents excessive water uptake by osmosis
- high pressure in plant cells (turgor pressure) makes the cell almost rigid, helping to support the plant.

2. Glycoproteins

Many animal cells secrete glycoproteins, consisting of a protein to which carbohydrate is attached. This forms an extracellular matrix. Tissues that consist of a single layer of cells produce a thin layer of extracellular matrix called the basement membrane, for example around blood capillaries and around alveoli in the lungs. The matrix is a gel and has these roles:
- supporting single layers of thin cells, which might otherwise tear or perforate
- cell to cell adhesion, for example, a basement membrane helps capillary wall cells to adhere to alveolus wall cells.

Cell division

THE CELL CYCLE IN EUKARYOTES

New cells are produced by division of existing cells. If many new cells are needed, cells go through a cycle of events again and again. This is called the cell cycle. The longest phase in this cycle is **interphase**. This is a very active period, during which the cell carries out many biochemical reactions and grows larger. The DNA molecules in the chromosomes are uncoiled and the genes on them can be transcribed, allowing the protein synthesis that is needed for growth. There is an increase in the number of mitochondria and in plant cells in the number of chloroplasts. There are three stages in interphase:

G_1 – a period of growth, DNA transcription and protein synthesis
S phase – the period during which all DNA in the nucleus is replicated
G_2 – a period in which the cell prepares for division.

At the end of interphase, the cell begins **mitosis** – the process by which the nucleus divides to form two genetically identical nuclei. Towards the end of mitosis, the cytoplasm of the cell starts to divide and eventually two cells are formed, each containing one nucleus. The process of dividing the cytoplasm to form two cells is **cytokinesis**. The two cells begin interphase when mitosis and cytokinesis have been completed.

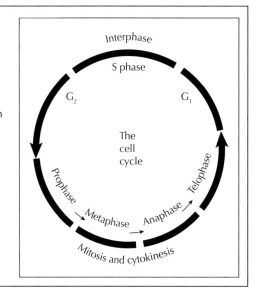

THE PHASES OF MITOSIS

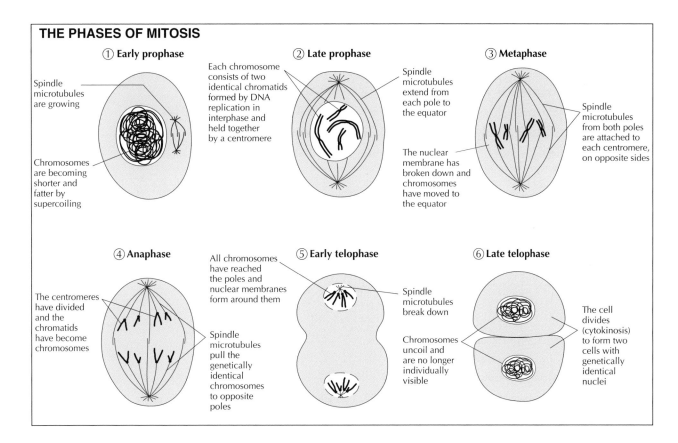

① **Early prophase**

Spindle microtubules are growing

Chromosomes are becoming shorter and fatter by supercoiling

② **Late prophase**

Each chromosome consists of two identical chromatids formed by DNA replication in interphase and held together by a centromere

Spindle microtubules extend from each pole to the equator

The nuclear membrane has broken down and chromosomes have moved to the equator

③ **Metaphase**

Spindle microtubules from both poles are attached to each centromere, on opposite sides

④ **Anaphase**

The centromeres have divided and the chromatids have become chromosomes

All chromosomes have reached the poles and nuclear membranes form around them

Spindle microtubules pull the genetically identical chromosomes to opposite poles

⑤ **Early telophase**

Spindle microtubules break down

Chromosomes uncoil and are no longer individually visible

⑥ **Late telophase**

The cell divides (cytokinosis) to form two cells with genetically identical nuclei

USES OF MITOSIS

Mitosis is used in eukaryotes whenever genetically identical cells are needed:

• during growth
• during embryonic development, when the large cell produced by fertilization (zygote) divides repeatedly to produce many smaller cells
• when tissues have been damaged and need to be repaired
• to reproduce asexually.

TUMOUR FORMATION

Sometimes the normal control of mitosis in a cell fails, due to a change in the genes of the cell. This cell divides into two, which inherit the change in the genes. The two daughter cells divide to form four cells. Repeated uncontrolled divisions soon produce a mass of cells called a tumour. This can happen in any tissue and in any organ. Tumours can grow to a large size and can spread to other parts of the body. The diseases caused by the growth of tumours are called cancer.

Cells 11

EXAM QUESTIONS ON TOPICS 1 AND 2

1 The photomicrograph below shows a transverse section of part of a liver cell.

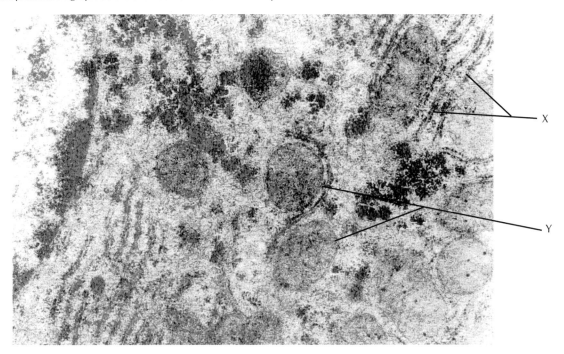

a) Identify the organelles labelled X and Y. [2]

b) On the photomicrograph, identify the nuclear membrane and show its position with a clear label. [1]

c) The liver cell shown in the photomicrograph was making large amounts of two substances.

Deduce what the two substances were, giving reasons for your answer based on the organelles visible in the photomicrograph. [2]

2 The diagram below represents the fluid model of a cell membrane.

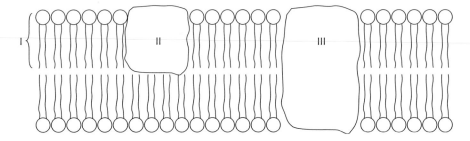

a) (i) State the name of the molecule labelled I. [1]

(ii) Label the diagram to show which part of molecule I is hydrophobic and which part is hydrophilic. [1]

b) (i) Identify whether molecule II is an integral or a peripheral protein. [1]

(ii) Describe the part played by molecule III in active transport. [2]

3 Ten teenage boys, aged 17 or 18, estimated their body fat percentage by measurements of skin fold thickness.

The estimates (%) were: 25.6, 12.9, 8.1, 10.2, 10.0, 8.9, 8.1, 15.3, 11.2, 13.7.

a) (i) Calculate the mean estimated body fat percentage. [2]

(ii) Calculate the standard deviation. [2]

The boys also measured their blood pressure. The boys whose estimated body fat percentages were higher tended to have higher blood pressure.

b) (i) What is this type of relationship between two variables called? [2]

(ii) Discuss whether this relationship proves that becoming obese causes high blood pressure. [2]

Water

POLARITY OF WATER

Water molecules consist of two hydrogen atoms bonded to an oxygen atom. The hydrogen atoms have a slight positive charge and the oxygen atom has a slight negative charge. So, water molecules have two poles – a positive hydrogen pole and a negative oxygen pole (below). This feature of a molecule is called **polarity**.

Water molecule

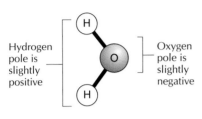

Hydrogen pole is slightly positive

Oxygen pole is slightly negative

HYDROGEN BONDING IN WATER

A bond can form between the positive pole of one water molecule and the negative pole of another. This is called a **hydrogen bond**. In liquid water many of these bonds form, giving water properties that make it a very useful substance for living organisms. The diagram (below) shows a hydrogen bond between two water molecules.

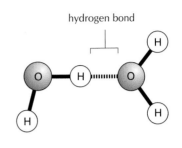

hydrogen bond

THE PROPERTIES OF WATER

Name of the property	Outline of the properties of water	Relationship between the properties of water and its uses in living organisms
Cohesion	Water molecules stick to each other because of the hydrogen bonds that form between them.	Strong pulling forces can be exerted to suck columns of water up to the tops of the tallest trees in their transport systems. These columns of water rarely break. Water is used as a **transport medium** in the xylem of plants.
Solvent properties	Many different substances dissolve in water because of its polarity (below). Inorganic particles with positive or negative charges dissolve, for example sodium ions. Organic substances with polar molecules dissolve, for example glucose. Enzymes also dissolve in water.	Most chemical reactions in living organisms take place with all of the substances involved in the reactions dissolved in water. Water is the **medium for metabolic reactions**. The solvent properties of water allow many substances to be carried dissolved in water in the blood of animals and the sap of plants. Water can be used as a **transport medium**.
Thermal properties: heat capacity	Water has a large heat capacity – large amounts of energy are needed to raise its temperature. The energy is needed to break some of the hydrogen bonds.	Blood, which is mainly composed of water, can carry heat from warmer parts of the body to cooler parts. Blood is used as a **transport medium** for heat.
Thermal properties: boiling point	The boiling point of water (100ºC) is high, because to change it from a liquid to a gas all of the hydrogen bonds between the water molecules have to be broken.	Water is below boiling point almost everywhere on Earth, and in most areas it is above freezing point. As a liquid, rather than a solid or a gas, it can act as the **medium for metabolic reactions**.
Thermal properties: the cooling effect of evaporation	Water can evaporate at temperatures below boiling point. Hydrogen bonds have to be broken to do this. The heat energy needed to break the bonds is taken from the liquid water, cooling it down.	Evaporation of water from plant leaves (transpiration) and from the human skin (sweat) has useful cooling effects. Water can be used as a **coolant**.

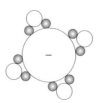

Ions with positive or negative charges dissolve as they are attracted to the negative or positive poles of water molecules.

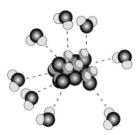

Many molecules are polar so are attracted to water molecules and dissolve.

Elements and compounds in living organisms

ELEMENTS IN LIVING ORGANISMS

Living organisms contain many chemical elements, some in large quantities and some in very small amounts. The four commonest chemical elements of life are carbon, hydrogen, oxygen and nitrogen. They are part of all the main organic compounds in living organisms. Examples of other elements that are needed are shown in the table opposite.

ORGANIC AND INORGANIC COMPOUNDS

Living organisms contain many chemical compounds. Some of them are organic and some are inorganic. Organic compounds are defined as compounds containing carbon that are found in living organisms. There are a few carbon compounds that are inorganic even though they can be found in living organisms. These are all simple carbon compounds that are also widely found in the environment. Carbon dioxide, carbonates and hydrogen carbonates are three examples of inorganic carbon compounds. All compounds that contain no carbon are inorganic. Three types of organic compound are found in large amounts in living organisms – carbohydrates, lipids and proteins.

SUBUNITS OF ORGANIC MACROMOLECULES

The molecules of many organic compounds are large and so are called macromolecules.
They are built up using small and relatively simple subunits. Some important subunits are shown below.

CHEMICAL ELEMENTS AND THEIR ROLES

Element and symbol	Role in plants, animals and prokaryotes
Sulphur S	Needed to make two of the twenty amino acids that proteins contain
Calcium Ca	Acts as a messenger, binding to calmodulin and other proteins that regulate processes inside cells, including transcription
Phosphorus P	Part of the phosphate groups in ATP and DNA molecules
Iron Fe	Needed to make cytochromes –proteins used for electron transport during aerobic cell respiration
Sodium Na	Pumped into the cytoplasm to raise the solute concentration and cause water to enter by osmosis

These elements have other specific roles in some organisms. For example, iron is needed to make hemoglobin in many animals and calcium is needed to make the minerals that strengthen bones and teeth.

Subunits of proteins, carbohydrates and lipids

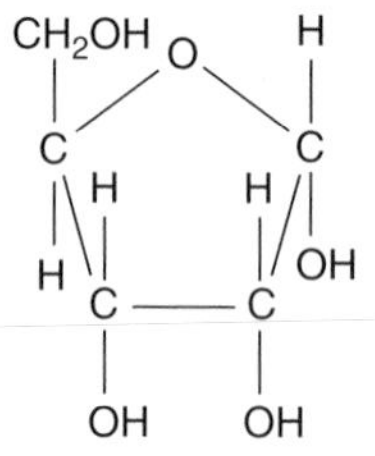

ribose
(a monosaccharide)

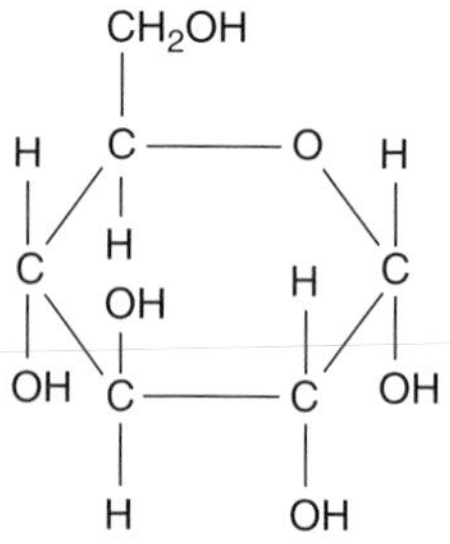

glucose
(a monosaccharide)

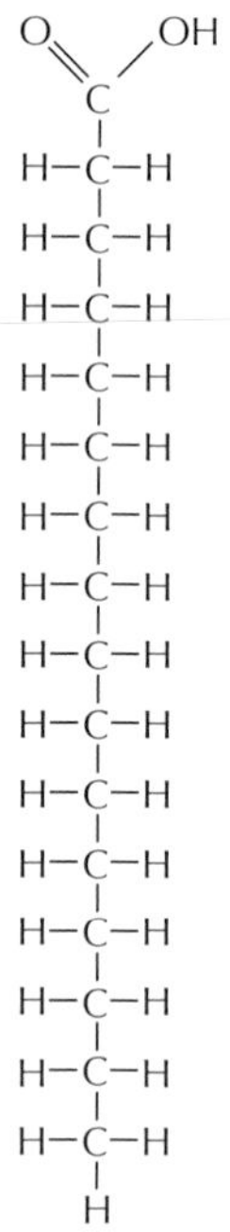

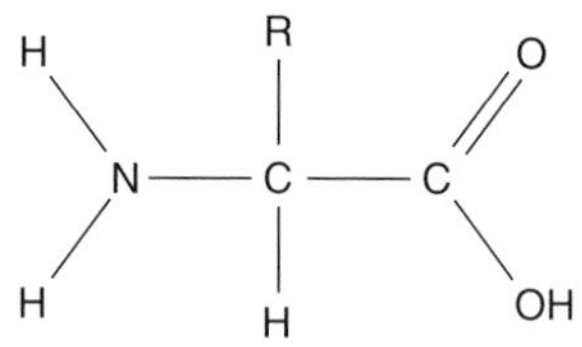

amino acids
(each of the twenty
amino acids in proteins
has a different R group)

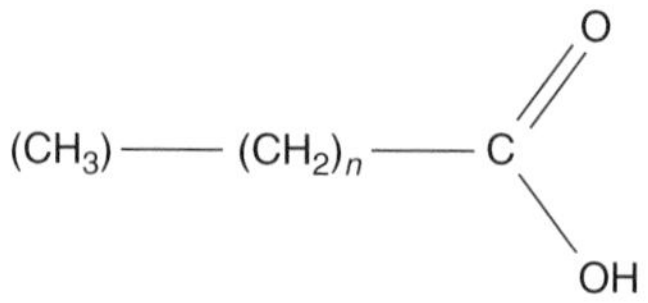

fatty acids
(general structure)

$(CH_3) - (CH_2)_n - C$

fatty acid
(number of carbon
atoms and bonding
between carbon
atoms varies)

Building macromolecules

CONDENSATION REACTIONS

In a condensation reaction two molecules are joined together to form a larger molecule. Water is also formed in the reaction. For example, two amino acids can be joined together to form a dipeptide by a condensation reaction. The new bond formed is a **peptide linkage**.

Condensation of two amino acids to form a dipeptide and water

Further condensation reactions can link amino acids to either end of the dipeptide, eventually forming a chain of many amino acids. This is called a **polypeptide**.

In a similar way, condensation reactions can be used to build up carbohydrates and lipids. The basic subunits of carbohydrates are monosaccharides. Two monosaccharides can be linked to form a disaccharide and more monosaccharides can be linked to a disaccharide to form a large molecule called a **polysaccharide**. Fatty acids can be linked to glycerol by condensation reactions to produce lipids called glycerides. A maximum of three fatty acids can be linked to each glycerol, producing a **triglyceride**.

HYDROLYSIS REACTIONS

Large molecules such as polypeptides, polysaccharides and triglycerides can be broken down into smaller molecules by hydrolysis reactions. Water molecules are used up in hydrolysis reactions. Hydrolysis reactions are the reverse of condensation reactions.

polypeptides + water $\longrightarrow$ dipeptides or amino acids

polysaccharides + water $\longrightarrow$ disaccharides or monosaccharides

glycerides + water $\longrightarrow$ fatty acids + glycerol

EXAMPLES OF CARBOHYDRATES

	Examples	Example of use in animals	Example of use in plants
Monosaccharides	glucose galactose fructose	Glucose is carried by the blood to transport energy to cells throughout the body	Fructose is used to make fruits sweet-tasting, attracting animals to disperse seeds in the fruit
Disaccharides	maltose lactose sucrose	Lactose is the sugar in milk, that provides energy to young mammals until they are weaned	Sucrose is carried by phloem to transport energy to cells throughout the plant
Polysaccharides	starch glycogen cellulose	Glycogen is used as a short-term energy store in liver and in muscles	Cellulose is used to make strong fibres that are used to construct the plant cell wall

FUNCTIONS OF LIPIDS

- **Energy storage** – in the form of fat in humans and oil in plants
- **Heat insulation** – a layer of fat under the skin reduces heat loss
- **Buoyancy** – lipids are less dense than water so help animals to float

CARBOHYDRATES AND LIPIDS IN ENERGY STORAGE

Both lipids and carbohydrates have advantages as energy storage compounds in living organisms. Carbohydrates are usually used for energy storage over short periods and lipids for long-term storage.

Advantages of lipids

1. Lipids contain more energy per gram than carbohydrates so stores of lipid are lighter than stores of carbohydrate that contain the same amount of energy

2. Lipids are insoluble in water, so they do not cause problems with osmosis in cells

Advantages of carbohydrates

1. Carbohydrates are more easily digested than lipids so the energy stored by them can be released more rapidly

2. Carbohydrates are soluble in water so are easier to transport to and from the store

Introducing DNA

THE NUCLEOTIDE SUBUNITS OF DNA

Although DNA is the genetic material of living organisms and is therefore of immense importance, it is made of relatively simple subunits. These are called **nucleotides**. Each nucleotide consists of three parts – a sugar (called deoxyribose), a phosphate group and a base. In diagrams of DNA structure these are usually shown as pentagons, circles and rectangles, respectively. The figure (below) shows how the sugar, the phosphate and the base are linked up in a nucleotide.

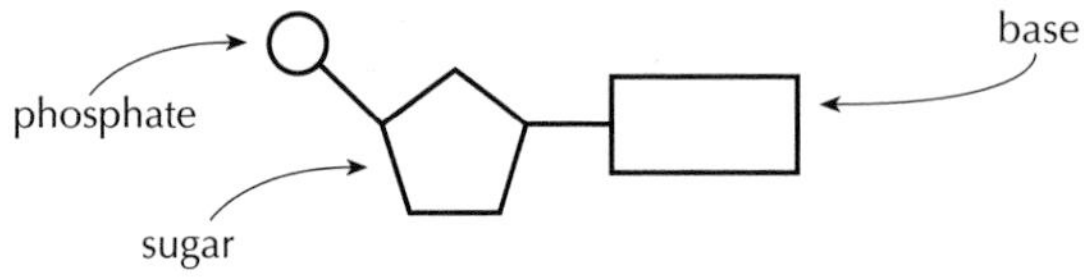

phosphate

sugar

base

DNA nucleotides do not all have the same base. Four different bases are found – adenine, cytosine, guanine and thymine. These are usually simply referred to as A, C, G and T.

BUILDING DNA MOLECULES

Two DNA nucleotides can be linked together by a covalent bond between the sugar of one nucleotide and the phosphate group of the other. More nucleotides can be added in a similar way to form a strand of nucleotides.

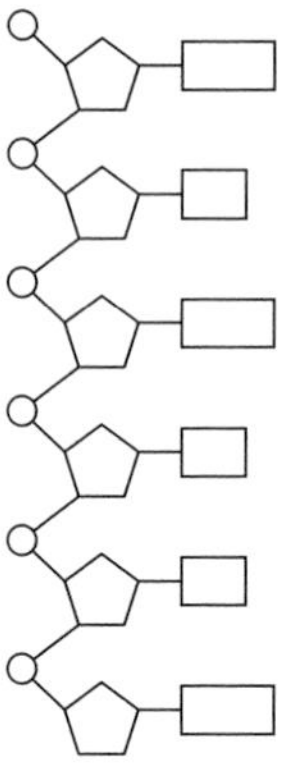

DNA molecules consist of two strands of nucleotides wound together into a double helix. Hydrogen bonds link the two strands together. These form between the bases of the two strands. However, adenine only forms hydrogen bonds with thymine and cytosine only forms hydrogen bonds with guanine. This is called **complementary base pairing**.

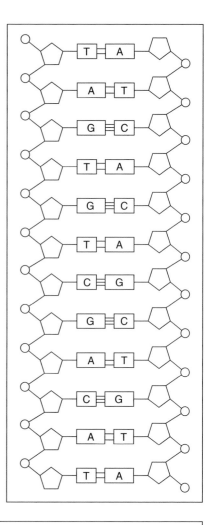

DNA REPLICATION

DNA replication is a way of copying DNA to produce new molecules with the same base sequence. It is **semi-conservative** – each molecule formed by replication consists of one new strand and one old strand conserved from the parent DNA molecule.

Stage 1
The DNA double helix is unwound and separated into strands by breaking the hydrogen bonds. Helicase is the main enzyme involved.

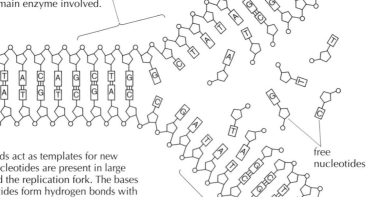

Stage 2
The single strands act as templates for new strands. Free nucleotides are present in large numbers around the replication fork. The bases of these nucleotides form hydrogen bonds with the bases on the parent strand. The nucleotides are linked up to form the new strand. DNA polymerase is the main enzyme involved.

Stage 3
The daughter DNA molecules each rewind into a double helix.

free nucleotides

The two daughter DNA molecules are identical in base sequence to each other and to the parent molecule, because of complementary base pairing (A pairs with T and C with G).
Each of the new strands is **complementary** to the template on which it was made and **identical** to the other template.

Transcription and translation

GENES AND POLYPEPTIDES

Polypeptides are long chains of amino acids. There are twenty different amino acids that can form part of a polypeptide. To make one particular polypeptide, amino acids must be linked up in a precise sequence. Genes store the information needed for making polypeptides. The information is stored in a coded form. The sequence of bases in a gene codes for the sequence of amino acids in a polypeptide.

The information in the gene is decoded during the making of the polypeptide. There are two stages in this process: **transcription** and **translation**.

DIFFERENCES BETWEEN DNA AND RNA

DNA and RNA both consist of chains of nucleotides, each composed of a sugar, a base and a phosphate. There are three differences between them.

Feature	DNA	RNA
Number of strands in the molecule	Two strands forming a double helix	One strand only
Type of sugar in each nucleotide	Deoxyribose	Ribose
Types of bases contained	A, C, G and T	A, C, G and U Uracil replaces thymine

TRANSCRIPTION

Instead of the DNA of genes being used directly to direct the synthesis of polypeptides, a copy is made. The copy is RNA. It carries the information needed to make a polypeptide out into the cytoplasm, so is called mRNA (messenger RNA). The copying of the base sequence of a gene by making an RNA molecule is called **transcription**. In transcription, the same rules of complementary base pairing are followed as in replication, except that uracil pairs with adenine, as RNA does not contain thymine. The RNA molecule produced therefore has a base sequence that is complementary to the transcribed strand and identical to the other DNA strand except that U replaces T.

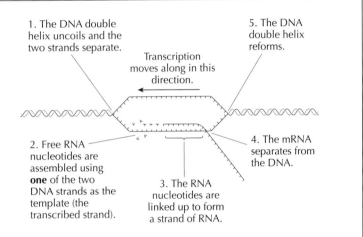

1. The DNA double helix uncoils and the two strands separate.

Transcription moves along in this direction.

5. The DNA double helix reforms.

2. Free RNA nucleotides are assembled using **one** of the two DNA strands as the template (the transcribed strand).

3. The RNA nucleotides are linked up to form a strand of RNA.

4. The mRNA separates from the DNA.

Stages 1, 2 and 3 are all carried out by the enzyme RNA polymerase.

TRANSLATION

Translation is carried out by ribosomes, using mRNA and tRNA. It is the **genetic code** that is being translated. The genetic code is a triplet code – three bases code for one amino acid. A group of three bases is called a **codon**.

1. Messenger RNA binds to the small subunit of the ribosome. The mRNA contains a series of **codons**, each of which codes for one amino acid.

2. Transfer RNA molecules are present around the ribosome in large numbers. Each tRNA has a special triplet of bases called an **anticodon** and carries the amino acid corresponding to this anticodon.

anticodon

large subunit of ribosome

amino acid

small subunit of ribosome

mRNA

4. The two amino acids carried by the tRNA molecules are bonded together by a peptide linkage. A dipeptide is formed, attached to the tRNA on the right. The tRNA on the left detaches. The ribosome moves along the mRNA to the next codon. Another tRNA carrying an amino acid binds. A chain of three amino acids is formed. These stages are repeated until a polypeptide is formed.

direction of movement of ribosome

3. tRNA molecules bind to the ribosome. Two can bind at once. tRNA can only bind if it has the anticodon that is complementary to the codon on the mRNA. The bases on the codon and anticodon link together by forming hydrogen bonds, following the same rules of complementary base pairing as in replication and transcription.

Genes, polypeptides and enzymes

ONE GENE–ONE POLYPEPTIDE HYPOTHESIS

Genes determine the amino acid sequence of proteins. However, some proteins contain more than one type of polypeptide. Hemoglobin is an example of this – it contains two different types of polypeptide. It was found that a different gene is needed to make each polypeptide. Further research has shown that there is almost always a single gene to code for a polypeptide, which does not code for any other polypeptide. This discovery led to an important hypothesis in molecular biology – the one gene–one polypeptide hypothesis. There are some exceptions to this general rule:

- Some genes code for transfer RNA or messenger RNA, not for polypeptides.
- Some DNA sequences act as regulators of gene expression and are not translated into polypeptides.
- In lymphocytes, pieces of DNA from different parts of the genome are spliced together and transcribed and translated to produce antibodies. Different lymphocytes produce different antibodies by splicing together DNA inherited from parents, in different ways.

INTRODUCING ENZYMES

Catalysts speed up chemical reactions without being changed themselves. Living organisms make biological catalysts called **enzymes**.

Enzymes are globular proteins which act as catalysts of chemical reactions.

Without enzymes to catalyse them, many chemical processes happen at a very slow rate in living organisms. By making some enzymes and not others, cells can control what chemical reactions happen in their cytoplasm.

The structure of enzymes is quite delicate and can be damaged by various substances and conditions. This is called **denaturation**.

Denaturation is changing the structure of an enzyme (or other protein) so that it can no longer carry out its function.

Denaturation is usually permanent.

In chemical reactions, one or more reactants are converted into one or more products. In reactions catalysed by enzymes, the reactants are called **substrates**.

ENZYME–SUBSTRATE SPECIFICITY

Most enzymes are specific – they catalyse very few different reactions. They therefore only have a very small number of possible substrates. This is called enzyme–substrate specificity. The substrates bind to a special region on the surface of the enzyme called the **active site**. *An active site is a region on the surface of an enzyme to which substrates bind and which catalyses a chemical reaction involving the substrates.*

The active site of an enzyme has a very intricate and precise shape. It also has distinctive chemical properties. Active sites match the shape and chemical properties of their substrates. Molecules of substrate fit the active site and are chemically attracted to it (right). Other molecules either do not fit or are not chemically attracted. They do not therefore bind to the active site. This is how enzymes are substrate-specific. The way in which the enzyme and substrate fit together is similar to the way in which a key fits a lock. The enzyme is like the lock and the substrate is like the key that fits it.

Stages in enzyme catalysis

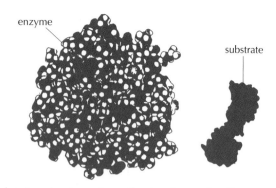

Substrate molecules are in continual random motion. If one collides with the active site it can bind to it.

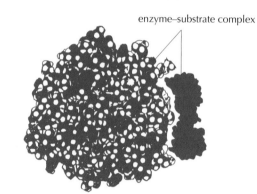

The substrate fits the active site. If other molecules collide with the active site they do not fit and fail to bind.

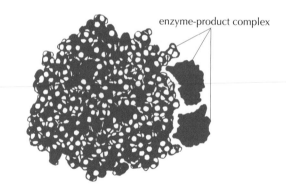

The active site catalyses a chemical reaction. The substrates are turned into products.

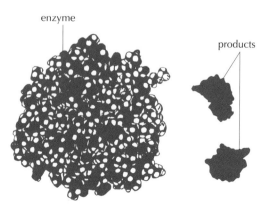

The products detach from the active site, leaving it free for more substrate to bind.

Enzymes in action

FACTORS AFFECTING ENZYME ACTIVITY

Wherever enzymes are used, it is important that they have the conditions that they need to work effectively. Temperature, pH and substrate concentration all affect the rate at which enzymes catalyse chemical reactions. The figures (below and right) show the relationships between enzyme activity and substrate concentration, temperature and pH.

EFFECT OF TEMPERATURE

Enzyme activity increases as temperature increases, often doubling with every 10 ∞C rise. This is because collisions between substrate and active site happen more frequently at higher temperatures due to faster molecular motion.

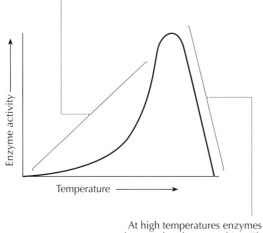

At high temperatures enzymes are denatured and stop working. This is because heat causes vibrations inside enzymes which break bonds needed to maintain the structure of the enzyme.

EFFECT OF pH

Optimum pH at which enzyme activity is fastest (pH 7 is optimum for most enzymes).

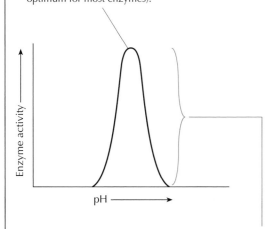

As pH increases or decreases from the optimum, enzyme activity is reduced. This is because the shape of the active site is altered so the substrate does not fit so well. Most enzymes are denatured by very high or low pH, so the enzyme no longer catalyses the reaction.

EFFECT OF SUBSTRATE CONCENTRATION

At low substrate concentrations, enzyme activity increases steeply as substrate concentration increases. This is because random collisions between substrate and active site happen more frequently with higher substrate concentrations.

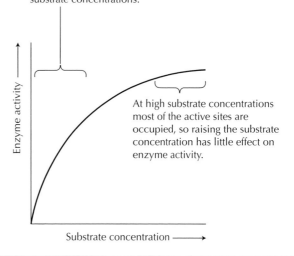

At high substrate concentrations most of the active sites are occupied, so raising the substrate concentration has little effect on enzyme activity.

LACTASE AND LACTOSE-FREE MILK

Lactose is the sugar that is naturally present in milk. It can be converted into glucose and galactose by the enzyme lactase.

$$lactose \xrightarrow{lactase} glucose + galactose$$

Lactase is obtained from *Kluveromyces lactis,* a type of yeast that grows naturally in milk. Biotechnology companies culture the yeast, extract the lactase from the yeast and purify it, for sale to food manufacturing companies. There are several reasons for using lactase in food processing:

- Some people are lactose intolerant and cannot drink more than about 250 ml of milk per day unless it is lactose-reduced.

- Galactose and glucose are sweeter than lactose, so less sugar needs to be added to sweet foods containing milk, such as milk shakes or fruit yoghurt.

- Lactose tends to crystallize during production of ice cream, giving a gritty texture. Because glucose and galactose are more soluble than lactose they remain dissolved, giving a smoother texture.

- Bacteria ferment glucose and galactose more quickly than lactose, so the production of yoghurt and cottage cheese is faster.

Lactase is used in two ways during food processing:

1. It can be added to milk. The final product contains the enzyme.

2. It can be immobilized on a surface or in beads of a porous material. The milk is then allowed to flow past the immobilised lactase. This avoids contamination of the product with lactase.

Cell respiration and energy

ENERGY AND CELLS

All living cells need a continual supply of energy. This energy is used for a wide range of processes including active transport and protein synthesis. Most of these processes require energy in the form of ATP (adenosine triphosphate). ATP is a chemical substance that can diffuse to any part of the cell and release energy.

Every cell produces its own ATP, by a process called **cell respiration**. In cell respiration, organic compounds such as glucose or fat are carefully broken down. Energy from them is used to make ATP. Cell respiration is defined as *controlled release of energy, in the form of ATP, from organic compounds in cells.*

Cell respiration can be aerobic or anaerobic. Aerobic cell respiration involves the use of oxygen and anaerobic cell respiration does not.

THE USE OF GLUCOSE IN RESPIRATION

Glucose is often the organic compound that is used in cell respiration. Chemical reactions in the cytoplasm break down glucose into a simpler organic compound called pyruvate. In these reactions a small amount of ATP is made using energy released from glucose.

ANAEROBIC CELL RESPIRATION

If no oxygen is available, the pyruvate remains in the cytoplasm and is converted into a waste product that can be removed from the cell. No ATP is produced in these reactions. In humans the waste product is lactate (lactic acid). In yeast the products are ethanol and carbon dioxide.

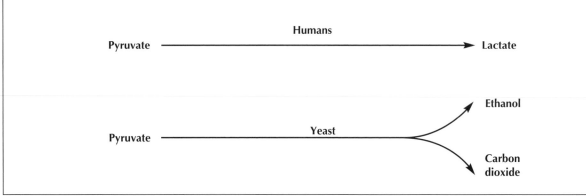

AEROBIC CELL RESPIRATION

If oxygen is available, the pyruvate is absorbed by the mitochondrion. Inside the mitochondrion the pyruvate is broken down into carbon dioxide and water. A large amount of ATP is produced as a result of these reactions. Aerobic cell respiration therefore has a much higher yield of ATP per gram of glucose than anaerobic cell respiration.

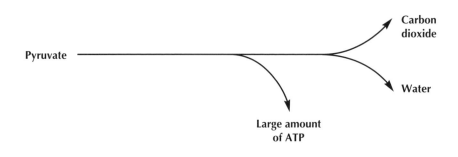

Photosynthesis

INTRODUCING PHOTOSYNTHESIS

Photosynthesis is the process used by plants and some other organisms to produce all their own organic substances (food), using only light energy and simple inorganic substances. It involves many stages and some complex chemical reactions, but it can be outlined in a series of statements.

- Photosynthesis involves an energy conversion. Light energy, usually sunlight, is converted into chemical energy.
- Sunlight is called white light, but it is actually made up of a wide range of wavelengths, including red, green and blue.
- Some substances called pigments can absorb light. The main pigment used to absorb light in photosynthesis is chlorophyll.
- The structure of chlorophyll allows it to absorb some colours or wavelengths of light better than others. Red and blue light are absorbed more than green.
- The green light that chlorophyll cannot absorb is reflected. This makes chlorophyll and therefore chloroplasts and plant leaves look green.
- Some of the energy absorbed by chlorophyll is used to produce ATP.
- Some of the energy absorbed by chlorophyll is used to split water molecules. This is called photolysis of water.
- Photolysis of water results in the formation of oxygen and hydrogen. The oxygen is released as a waste product.
- Carbon dioxide is absorbed for use in photosynthesis. The carbon from it is used to make a wide range of organic substances. The conversion of carbon in a gas to carbon in solid compounds is called carbon fixation.
- Carbon fixation involves the use of hydrogen from photolysis and energy from ATP.

MEASURING RATES OF PHOTOSYNTHESIS

Photosynthesis involves the production of oxygen, the uptake of carbon dioxide and an increase in biomass. Any of these can be used as a measure of the rate of photosynthesis.

Production of oxygen

Aquatic plants (e.g. *Myriophyllum*) release bubbles of oxygen when they carry out photosynthesis. If these bubbles are collected, their volume can be measured.

Uptake of carbon dioxide

Leaves take in CO_2 from the air or water around them, but this is difficult to measure directly. If CO_2 is absorbed from water, the pH of the water rises. This can be monitored with pH indicators or with pH meters.

Increases in biomass

If batches of plants are harvested at a series of times and the biomass of the batches is determined, the rate of increase in biomass gives an indirect measure of the rate of photosynthesis in the plants.

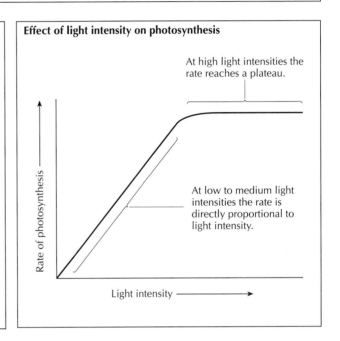

Effect of light intensity on photosynthesis

At high light intensities the rate reaches a plateau.

At low to medium light intensities the rate is directly proportional to light intensity.

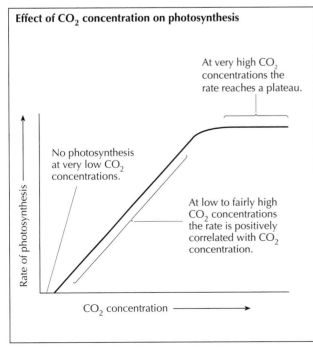

Effect of CO_2 concentration on photosynthesis

At very high CO_2 concentrations the rate reaches a plateau.

No photosynthesis at very low CO_2 concentrations.

At low to fairly high CO_2 concentrations the rate is positively correlated with CO_2 concentration.

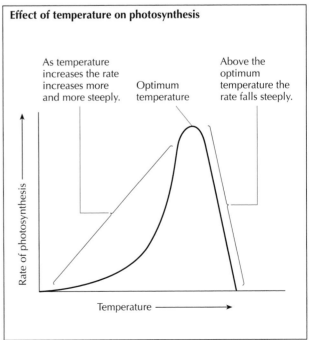

Effect of temperature on photosynthesis

As temperature increases the rate increases more and more steeply.

Optimum temperature

Above the optimum temperature the rate falls steeply.

1 The table below shows the base composition of genetic material from ten sources.

Source of genetic material	Base composition (%)				
	Adenine	Guanine	Thymine	Cytosine	Uracil
Cattle thymus gland	28.2	21.5	27.8	22.5	0.0
Cattle spleen	27.9	22.7	27.3	22.1	0.0
Cattle sperm	28.7	22.2	27.2	22.0	0.0
Pig thymus gland	30.0	20.4	28.9	20.7	0.0
Salmon	29.7	20.8	29.1	20.4	0.0
Wheat	27.3	22.7	27.1	22.8	0.0
Yeast	31.3	18.7	32.9	17.1	0.0
E. coli (bacteria)	26.0	24.9	23.9	25.2	0.0
human sperm	31.0	19.1	31.5	18.4	0.0
influenza virus	23.0	20.0	0.0	24.5	32.5

a) Deduce the type of genetic material used by

   (i) cattle   [1]

   (ii) *E. coli*   [1]

   (iii) influenza viruses.   [1]

b) Suggest a reason for the difference between cattle thymus gland, spleen and sperm in the measurements of their base composition.   [1]

c) (i) Explain the reasons for the total amount of adenine plus guanine being close to 50% in the genetic material of many of the species in the table.   [3]

   (ii) Identify two other trends in the base composition of the species that have 50% adenine and guanine.   [2]

d) (i) Identify a species shown in the table that does not follow the trends in base composition described in (c).   [1]

   (ii) Explain the reasons for the base composition of this species being different.   [2]

2 The graph (right) shows the results of a data logging experiment. *Chlorella*, a type of alga that is often used in photosynthesis experiments, was cultured in water in a large glass vessel. Light intensity, temperature and the pH of the water were monitored over a three-day period. The changes in pH were due to changes in carbon dioxide concentration. An increase in CO_2 concentration causes a decrease in pH.

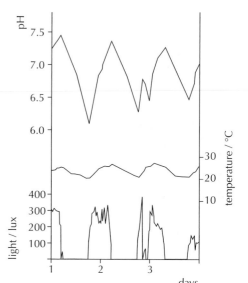

a) State the relationship shown in the graph between

   (i) light intensity and CO_2 concentration   [1]

   (ii) temperature and CO_2 concentration.   [1]

b) Deduce, from the data in the graph, whether the effect of light intensity or temperature on carbon dioxide concentration is greater.   [2]

c) The graph shows both rises and falls in CO_2 concentration. Explain the causes of

   (i) rises in CO_2 concentration   [2]

   (ii) falls in CO_2 concentration.   [2]

3 The diagram shows the basic structure of amino acids

a) State what is represented in the diagram by the letter R.   [1]

b) Draw a simple diagram to show how two amino acids are linked together.   [2]

c) Amino acids are linked together to form polypeptides at special sites in the cytoplasm of both prokaryotic and eukaryotic cells. Compare the sites where polypeptides are formed in prokaryotic cells with the sites in eukaryotic cells.   [2]

Genes and chromosomes

GENES

Genetics is the study of variation and inheritance. The basic unit of inheritance is the **gene**. *A gene is a heritable factor that controls a specific characteristic.*

A typical animal or plant cell nucleus contains thousands of genes. The total number of genes in humans is not yet known but is probably between 30 000 and 40 000. All of the genes of an organism are known collectively as the **genome**. *A genome is the whole of the genetic information of an organism.*

CHROMOSOMES

Genes are made of DNA. They are part of much larger DNA molecules called **chromosomes**. In eukaryotes, proteins are always associated with the DNA in chromosomes.

A typical animal or plant chromosome contains about a 1000 genes, which are arranged in a linear sequence. In any particular type of chromosome the same genes are found arranged in the same sequence. The position of a gene on a chromosome is called the **gene locus**.

ALLELES

Although one particular chromosome type always has the same genes in the same sequence, the genes themselves can vary. Different forms of many genes can be found. These are called **alleles** of the gene. *An allele is a form of a gene, differing from other alleles of the gene by a few bases at most and occupying the same locus as the other alleles of that gene.*

REPLICATION OF CHROMOSOMES

If a nucleus is going to divide by mitosis or meiosis, all DNA in the nucleus is replicated. When mitosis or meiosis begins, each chromosome is visible as a double structure (see below). The two parts are called **chromatids** and are connected by a centromere. Some types of chromosome have a **centromere** in the centre and others have a centromere nearer to one end.

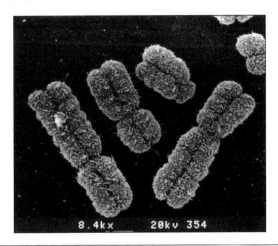

GENE MUTATION

Genes are almost always passed from parent to offspring without being changed. Occasionally genes do change and this is called **gene mutation**.

Gene mutation is a change to the base sequence of a gene. The smallest possible change is when one base in a gene is replaced by another base. This type of gene mutation is called a **base substitution**. Although only one base is changed, the consequences can be very significant. Many gene mutations cause a genetic disease. More than four thousand genetic diseases have been discovered in humans. One example is sickle cell anemia (right).

Sickle cell anemia – the consequences of a base substitution mutation

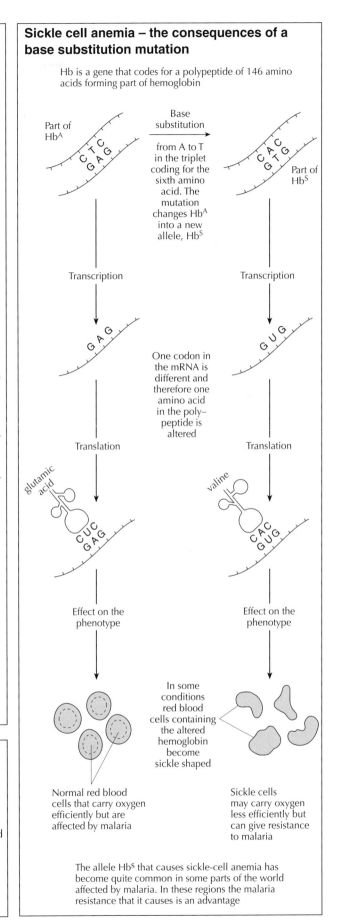

Hb is a gene that codes for a polypeptide of 146 amino acids forming part of hemoglobin

Part of Hb^A

Base substitution

from A to T in the triplet coding for the sixth amino acid. The mutation changes Hb^A into a new allele, Hb^S

Part of Hb^S

Transcription

Transcription

One codon in the mRNA is different and therefore one amino acid in the poly– peptide is altered

Translation

Translation

glutamic acid

valine

Effect on the phenotype

Effect on the phenotype

In some conditions red blood cells containing the altered hemoglobin become sickle shaped

Normal red blood cells that carry oxygen efficiently but are affected by malaria

Sickle cells may carry oxygen less efficiently but can give resistance to malaria

The allele Hb^S that causes sickle-cell anemia has become quite common in some parts of the world affected by malaria. In these regions the malaria resistance that it causes is an advantage

Meiosis

HAPLOID AND DIPLOID

In most cells the nucleus contains two of each type of chromosome (right). The cell therefore has two full sets of chromosomes. This is called **diploid**. Some cells only contain one of each type of chromosome and therefore have just one set. This is called **haploid**.

In diploid cells each pair of chromosomes have the same genes, arranged in the same sequence. However, they do not usually have the same alleles of all of these genes. They are therefore not identical but instead are **homologous**.

Homologous chromosomes have the same genes as each other, in the same sequence, but not necessarily the same alleles of those genes.

The number of chromosomes in a cell can be reduced from diploid to haploid by the process of meiosis. Meiosis is described as a **reduction division**. Living organisms that reproduce sexually have to halve their chromosome number at some stage in the life cycle because the fusion of gametes during fertilization doubles it.

Chromosomes of a human female

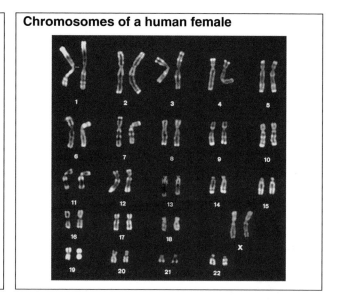

STAGES OF MEIOSIS

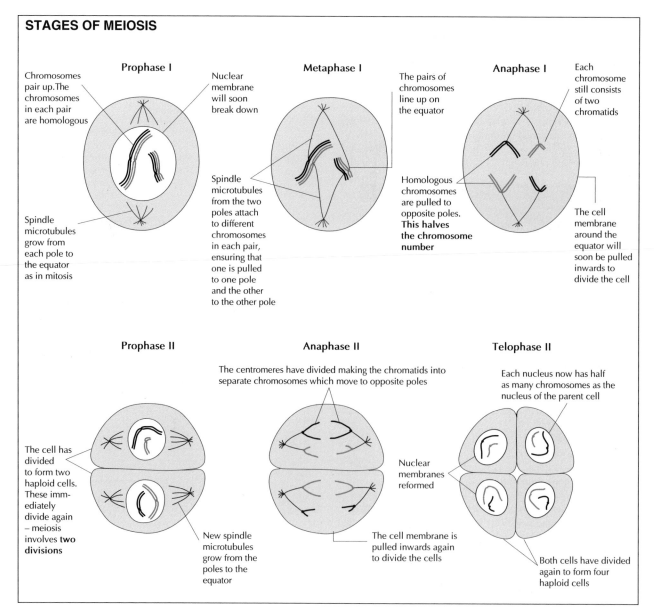

Prophase I

Chromosomes pair up. The chromosomes in each pair are homologous

Nuclear membrane will soon break down

Spindle microtubules grow from each pole to the equator as in mitosis

Metaphase I

Spindle microtubules from the two poles attach to different chromosomes in each pair, ensuring that one is pulled to one pole and the other to the other pole

Anaphase I

The pairs of chromosomes line up on the equator

Homologous chromosomes are pulled to opposite poles. **This halves the chromosome number**

Each chromosome still consists of two chromatids

The cell membrane around the equator will soon be pulled inwards to divide the cell

Prophase II

The cell has divided to form two haploid cells. These immediately divide again – meiosis involves **two divisions**

New spindle microtubules grow from the poles to the equator

Anaphase II

The centromeres have divided making the chromatids into separate chromosomes which move to opposite poles

Nuclear membranes reformed

The cell membrane is pulled inwards again to divide the cells

Telophase II

Each nucleus now has half as many chromosomes as the nucleus of the parent cell

Both cells have divided again to form four haploid cells

Karyotypes

KARYOTYPES AND KARYOTYPING

The number and appearance of the chromosomes in an organism is called the **karyotype**. Living organisms that are members of the same species usually have the same karyotype. The karyotype of a human female is shown on page 24.

From a karyotype, the gender of a person can be deduced and chromosome abnormalities can be detected. The most useful time to do this is before birth. Cells have to be obtained from the fetus. There are two ways of doing this:

1. Amniocentesis

A sample of amniotic fluid is removed from the amniotic sac around the fetus. To do this, a hypodermic needle is inserted through the wall of the mother's abdomen and wall of the uterus. Amniotic fluid is drawn out into a syringe. It contains cells from the fetus.

2. Chorionic villus sampling

Cells are removed from fetal tissues in the placenta called chorionic villi. As with amniocentesis a hypodermic needle, inserted through the mother's abdomen and uterus wall, is used to obtain the cells.

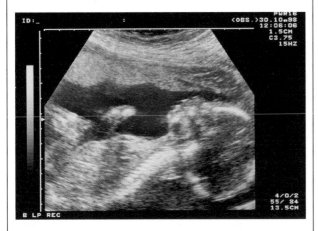

Once fetal cells have been obtained, they are incubated with chemicals that stimulate them to divide by mitosis. Another chemical is used which stops mitosis in metaphase of mitosis. Chromosomes are most easily visible in metaphase. A fluid is used to burst the cells and spread out the chromosomes. The burst cells are examined using a microscope and a photograph is taken of the chromosomes from one cell (below). The chromosomes in the photograph are cut out and arranged into pairs according to their size and structure. This is called **karyotyping**.

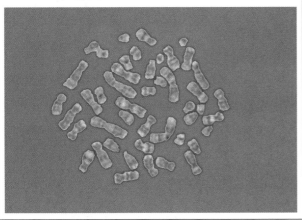

ANALYSIS OF KARYOTYPES

The gender of the fetus can be determined from the sex chromosomes. Gender determination is described on page 28. Karyotypes can also be analysed to find out whether a fetus has chromosome abnormalities. Sometimes chromosomes that should separate and move to opposite poles during meiosis do not separate and instead move to the same pole. This can happen in either the first (below left) or the second division of meiosis (below right). Non-separation of chromosomes is called **non-disjunction**. The result is that gametes are produced with either one chromosome too many or too few.

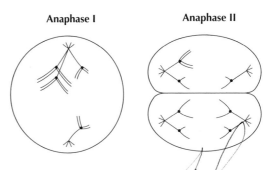

Anaphase I   Anaphase II

Gametes with one chromosome too few usually quickly die. Gametes with one chromosome too many sometimes survive. When they are fertilised, a zygote is produced with three chromosomes of one type instead of two. For example, in the karyotype (below) there are 47 chromosomes in total with three chromosomes of type 21, rather than two. This causes Down syndrome. It can be due either to non-disjunction during the formation of the sperm or egg. The chance of Down syndrome increases with the age of the parents.

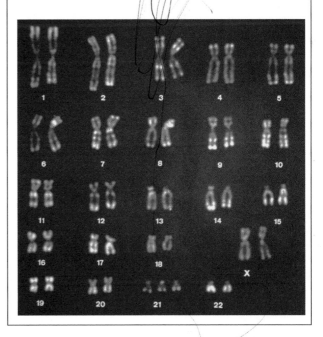

Monohybrid crosses

MENDEL'S MONOHYBRID CROSSES

Gregor Mendel is often regarded as the father of genetics. He investigated inheritance by crossing varieties of pea plants that had different characteristics. For example, he crossed a variety that had round seeds with a variety that had wrinkled seeds. He found that all the offspring (called the **F_1 generation**) had the same characteristic as one of the parents. He allowed the F_1 generation to self-fertilize – each plant produced offspring by fertilizing its female gametes with its own male gametes. The offspring (called the **F_2 generation**) contained both of the original parental types. The characteristic that disappeared in the F_1 generation reappeared in a quarter of the F_2 generation. Mendel deduced that inheritance is based on factors that can be passed on from generation to generation. We now call these factors **genes**. Different forms of a gene are called **alleles**. The figure below shows an example of Mendel's monohybrid crosses.

DEFINITIONS OF TERMS USED IN GENETICS

There are two pairs of terms that are often used by geneticists:
• **Homozygous** – *having two identical alleles of a gene.* All the gametes of a homozygote have the same allele.
• **Heterozygous** – *having two different alleles of a gene.* Half of the gametes of a heterozygote have one of the alleles and half have the other allele.
• **Dominant allele** – *an allele that has the same effect on the phenotype in a heterozygous individual (where it is combined with a recessive allele) as in a homozygous individual (where there are two copies of the dominant allele).*
• **Recessive allele** – *an allele that only has an effect on the phenotype in homozygous individuals (where there are two copies of the recessive allele).* In heterozygous individuals the recessive allele is hidden by the dominant allele.

Monohybrid cross between smooth and wrinkled seed pea plants

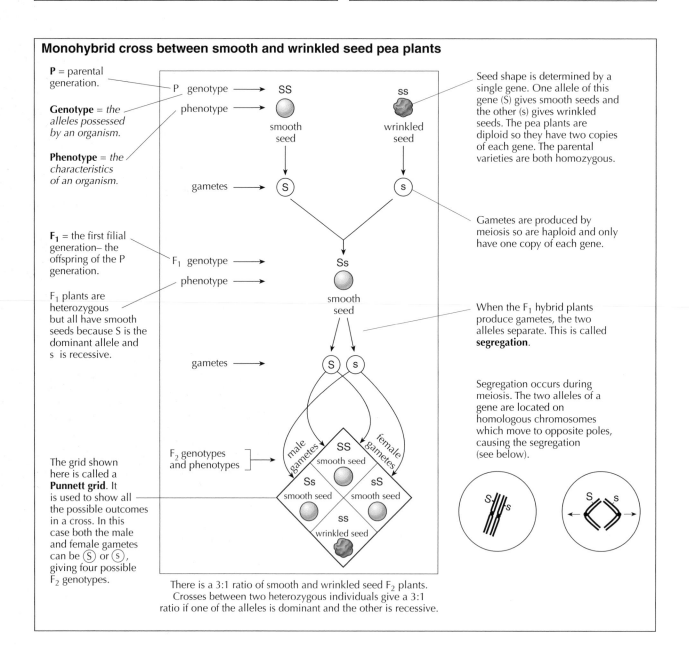

P = parental generation.

Genotype = *the alleles possessed by an organism.*

Phenotype = *the characteristics of an organism.*

F_1 = the first filial generation– the offspring of the P generation.

F_1 plants are heterozygous but all have smooth seeds because S is the dominant allele and s is recessive.

The grid shown here is called a **Punnett grid**. It is used to show all the possible outcomes in a cross. In this case both the male and female gametes can be Ⓢ or Ⓢ, giving four possible F_2 genotypes.

Seed shape is determined by a single gene. One allele of this gene (S) gives smooth seeds and the other (s) gives wrinkled seeds. The pea plants are diploid so they have two copies of each gene. The parental varieties are both homozygous.

Gametes are produced by meiosis so are haploid and only have one copy of each gene.

When the F_1 hybrid plants produce gametes, the two alleles separate. This is called **segregation**.

Segregation occurs during meiosis. The two alleles of a gene are located on homologous chromosomes which move to opposite poles, causing the segregation (see below).

There is a 3:1 ratio of smooth and wrinkled seed F_2 plants. Crosses between two heterozygous individuals give a 3:1 ratio if one of the alleles is dominant and the other is recessive.

Inheritance of blood groups

The principles of inheritance discovered by Mendel in pea plants also operate in other plants and in animals. There are, however, sometimes differences and two of these are demonstrated by the inheritance of ABO blood groups in humans – codominance and multiple alleles.

CROSS INVOLVING CODOMINANT ALLELES

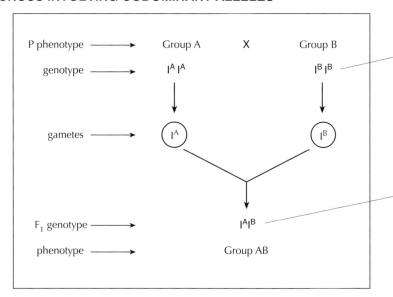

I^A is the allele for blood group A and I^B is the allele for blood group B. Neither allele is recessive, so both are given upper case letters as their symbol.

If I^A and I^B are present together, they both affect the phenotype because they are codominant. Codominant alleles are pairs of alleles that both affect the phenotype when present together in a heterozygote.

CROSS INVOLVING MULTIPLE ALLELES

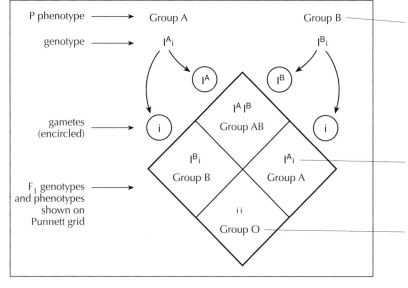

The gene that controls **ABO** blood groups has a third allele: i If there are more than two alleles of a gene, they are called *multiple alleles.*

i is recessive to both I^A and I^B so I^A i gives blood group **A** and I^B i gives blood group B.

Individuals who are homozygous for i are in blood group O.

DEDUCING GENOTYPES FROM PEDIGREE CHARTS

A pedigree chart shows the members of a family and how they are related to each other. Males are shown as squares and females as circles. If the phenotypes of the members of the family are known, the genotypes can often be deduced. The figure (right) is a pedigree chart that shows the blood group of each individual. All of the genotypes can be deduced. It is also possible to deduce the probability of the first child of the parents in the third generation being blood group A, B, AB and O.

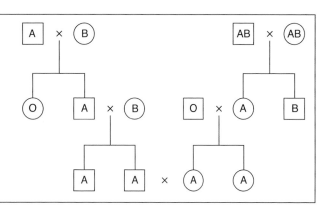

Genes and gender

SEX CHROMOSOMES AND GENDER

- Two chromosomes determine the gender of a child (whether it is male or female). These are called the sex chromosomes.
- The X chromosome is relatively large and carries many genes.
- The Y chromosome is much smaller and carries only a few genes.
- If two X chromosomes are present in a human embryo and no Y chromosome, it develops into a girl.
- If one X chromosome and one Y chromosome are present, a human embryo develops into a boy.
- When women reproduce, they pass on one X chromosome in the egg.
- When men reproduce, they pass on either one X or one Y chromosome in the sperm, so the gender of a child depends on whether the sperm that fertilizes the egg is carrying an X or a Y chromosome (right).

Inheritance of gender in humans

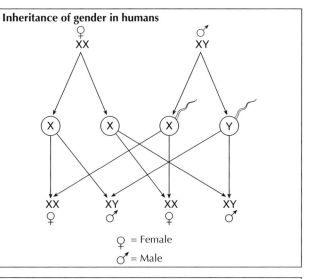

$\female$ = Female
$\male$ = Male

SEX LINKAGE

If a gene is carried on the X chromosome, the pattern of inheritance is different for males and females – there is sex linkage. *Sex linkage is the association of a characteristic with gender, because the gene controlling the characteristic is located on a sex chromosome.* Sex-linked genes are almost always located on the X chromosome. Females have two X chromosomes and therefore have two copies of sex linked genes. Males only have one X chromosome and therefore only have one copy of sex linked genes. In humans, hemophilia (below) and red–green colour blindness are examples of sex-linked characteristics.

CHOOSING SYMBOLS FOR ALLELES

These rules are usually followed when choosing symbols for alleles:

1. **One dominant and one recessive allele of a gene**
   A letter is chosen. The dominant allele is shown with the upper case, and the recessive allele with the lower case letter (e.g. A and a)
2. **Co-dominant alleles**
   A letter is chosen. This letter and a superscript letter represent each allele. (e.g. C^w and C^r)
3. **Sex-linked dominant and recessive alleles**
   The letter X is used to show the X chromosome. Each allele is shown superscript. (e.g. X^H and X^h)

Example of a cross involving sex linkage

The diagram below shows how two parents, neither of whom have hemophilia, could have a hemophiliac son.

The mother is heterozygous but is not a hemophiliac because H is dominant and h is recessive. She is a **carrier** of the allele for hemophilia.

A carrier has a recessive allele of a gene but it does not affect the phenotype because a dominant allele is also present.

KEY

X^H X chromosome carrying the allele for normal blood clotting

X^h X chromosome carrying the allele for hemophilia.

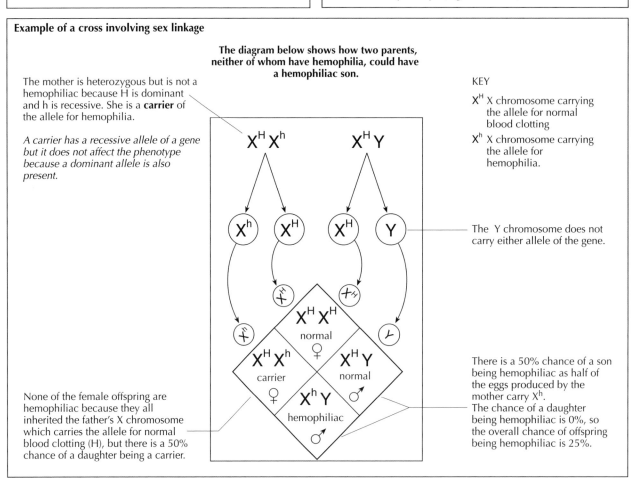

The Y chromosome does not carry either allele of the gene.

None of the female offspring are hemophiliac because they all inherited the father's X chromosome which carries the allele for normal blood clotting (H), but there is a 50% chance of a daughter being a carrier.

There is a 50% chance of a son being hemophiliac as half of the eggs produced by the mother carry X^h.
The chance of a daughter being hemophiliac is 0%, so the overall chance of offspring being hemophiliac is 25%.

Deducing genotypes

USING PEDIGREE CHARTS

Pedigree charts can be used to deduce whether a character is caused by a dominant or recessive allele and whether it is sex-linked or not. They can also be used to deduce the genotypes of individuals. The figures (below) are pedigree charts that each show a different pattern of inheritance. Squares represent males and circles represent females. Shaded symbols represent individuals affected by the condition and unshaded symbols represent unaffected individuals. The probability of the different phenotypes in the offspring of some of the couples in the pedigrees (marked with an asterisk *) can be determined.

MUSCULAR DYSTROPHY

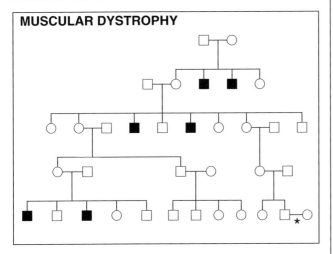

ALBINISM

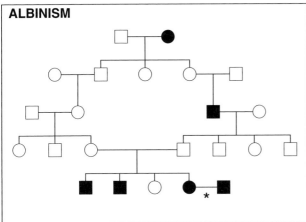

HUNTINGTON'S DISEASE

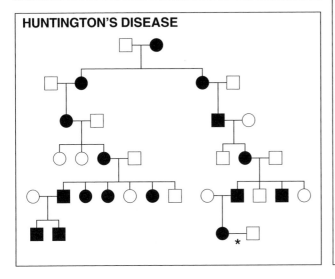

USING TEST CROSSES

It is not always possible to discover whether an individual has a gene, or does not have it, by looking at the individual's phenotype. If one allele of a gene is dominant and another allele is recessive, an individual with two copies of the dominant allele has the same phenotype as an individual with one dominant and one recessive allele. These two genotypes can be distinguished by carrying out a **test cross**.

In a test cross an individual that might be heterozygous is crossed with an individual that is homozygous recessive.

Example of a test cross

A farmer is unsure whether his bull is a purebred Hereford or whether it is a Hereford x Aberdeen Angus hybrid. Hereford cattle have a white head caused by a dominant allele (H). Aberdeen Angus cattle have black heads caused by a recessive allele of the same gene (h). The farmer crosses his bull with 100 Aberdeen Angus cows. The figures below show the possible outcomes.

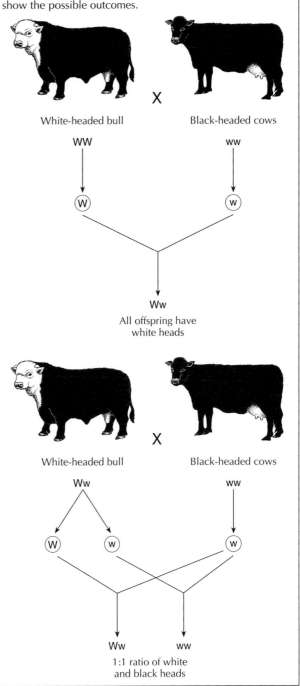

DNA profiling

PCR – POLYMERASE CHAIN REACTION

In the polymerase chain reaction, DNA is copied again and again to produce many copies of the original molecules. Millions of copies of the DNA can be produced in a few hours. This is very useful when very small quantities of DNA are found in a sample and larger amounts are needed for analysis. DNA from very small samples of semen, blood or other tissue or even from long-dead specimens can be amplified using PCR.

PCR is carried out at high temperatures using a DNA polymerase enzyme from *Thermus aquaticus*, a bacterium that lives in hot springs.

GEL ELECTROPHORESIS

Gel electrophoresis is a method of separating mixtures of proteins, DNA or other molecules that are charged. The mixture is placed on a thin sheet of gel, which acts like a molecular sieve. An electric field is applied to the gel by attaching electrodes to both ends. Depending on whether the particles are positively or negatively charged, they move towards one of the electrodes or the other. The rate of movement depends on the size and charge of the molecules – small and highly charged molecules move faster than larger or less charged ones.

DNA PROFILING

Humans and other organisms have short sequences of bases that are repeated many times called satellite DNA. This satellite DNA varies greatly between different individuals in the number of repeats. If it is copied using PCR and then cut up into short fragments using restriction enzymes, the lengths of the fragments vary greatly between individuals. Gel electrophoresis can be used to separate fragmented pieces of DNA according to their charge and size. The pattern of bands on the gel is very unlikely to be the same for any two individuals. This technique, called DNA profiling or DNA fingerprinting has many applications, including forensic investigations (obtaining evidence to use in court cases) and investigating paternity (who the father of a child is).

Forensic use of DNA profiling

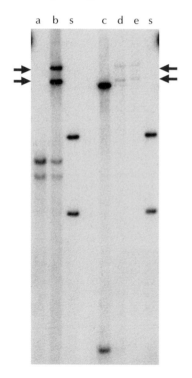

The first use of DNA profiling was in the Enderby double murder case. The DNA profiles show clearly whether the prime suspect was guilty.

- a = hair roots from the first victim
- b = mixed semen and vaginal fluids from the first victim
- c = blood of second victim
- d = vaginal swab from second victim
- e = semen stain on second victim
- s = blood of prime suspect.

The two bands indicated by arrows are from DNA in the culprit's semen.

Testing paternity using DNA profiling

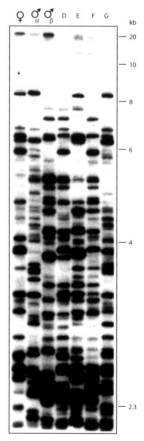

The DNA profiles of a family of dunnocks (*Prunella modularis*) are shown above. Dunnocks are small birds found in Europe, North Africa and Asia. The tracks from left to right are: the female, two resident males that might have been the father of the offspring and four offspring. The results show that the β male fathered three of the four offspring (D, E and F), despite being less dominant than the α male.

Genetic modification

GENETIC MODIFICATION AND ITS USES

The genetic code is universal, so genes can be transferred from one organism to another, even if they are members of different species. A gene codes for a polypeptide with the same amino acid sequence, whether it is in a human cell, a bacterium or any other cell. Organisms that have had genes transferred to them are called **genetically modified organisms** (GMO) or transgenic organisms. The process of transferring genes is called genetic modification. One example is the transfer of a gene for factor IX (a blood clotting factor) from humans to sheep, where it is produced in the sheep's milk. Another example is the transfer of the gene for resistance to the herbicide glyphosate from a bacterium to crop plants, so that the crop can be sprayed with the herbicide.

Techniques used for gene transfer into bacteria

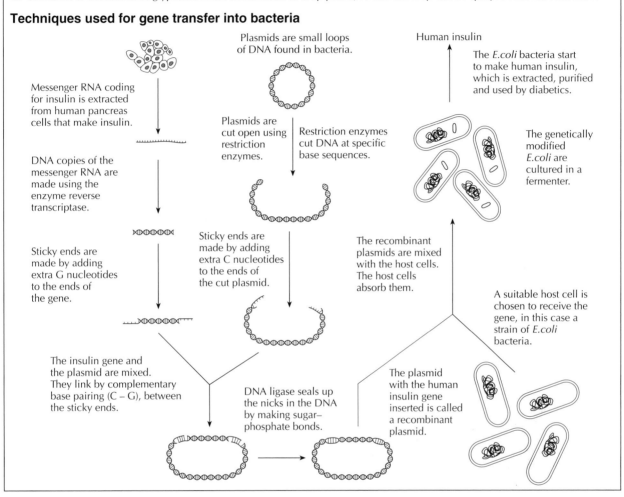

Plasmids are small loops of DNA found in bacteria.

Human insulin

The *E.coli* bacteria start to make human insulin, which is extracted, purified and used by diabetics.

Messenger RNA coding for insulin is extracted from human pancreas cells that make insulin.

Plasmids are cut open using restriction enzymes.

Restriction enzymes cut DNA at specific base sequences.

The genetically modified *E.coli* are cultured in a fermenter.

DNA copies of the messenger RNA are made using the enzyme reverse transcriptase.

Sticky ends are made by adding extra C nucleotides to the ends of the cut plasmid.

The recombinant plasmids are mixed with the host cells. The host cells absorb them.

A suitable host cell is chosen to receive the gene, in this case a strain of *E.coli* bacteria.

Sticky ends are made by adding extra G nucleotides to the ends of the gene.

The insulin gene and the plasmid are mixed. They link by complementary base pairing (C – G), between the sticky ends.

DNA ligase seals up the nicks in the DNA by making sugar–phosphate bonds.

The plasmid with the human insulin gene inserted is called a recombinant plasmid.

BENEFITS AND RISKS OF GENETIC MODIFICATION

The production of human insulin using bacteria has enormous benefits and no obvious harmful effects. There are other examples of genetic modification that are more controversial. Maize crops are often seriously damaged by corn borer insects. A gene from a bacterium (*Bacillus thuringiensis*) has been transferred to maize. The gene codes for a bacterial protein called Bt toxin that kills corn borers feeding on the maize.

Potential benefits of Bt maize	Possible harmful effects of Bt maize
1. Less pest damage and therefore higher crop yields to help to reduce food shortages	1. Humans or farm animals that eat the genetically modified maize might be harmed by the bacterial DNA in it, or by the Bt toxin
2. Less land needed for crop production, so some could become areas for wildlife conservation	2. Insects that are not pests could be killed. Maize pollen containing the toxin is blown onto wild plants growing near the maize. Insects feeding on the wild plants, including Monarch butterfly caterpillars, are therefore affected even if they do not feed on the maize
3. Less use of insecticide sprays, which are expensive and can be harmful to farm workers and to wildlife	3. Populations of wild plants might be changed. Cross-pollination will spread the Bt gene into some wild plants but not others. These plants would then produce the Bt toxin and have an advantage over other wild plants in the struggle for survival

Cloning

CLONES AND CLONING

Cloning is producing identical copies of genes, cells or organisms. The products of cloning are called a **clone**.
A clone is a group of genetically identical organisms or a group of genetically identical cells derived from a single parent cell.
Cloning is very useful if an organism has a desirable combination of characteristics and more organisms with the same characteristics are wanted – this is reproductive cloning. Sometimes cloning is used to produce skin or other tissues needed to treat a patient – this is therapeutic cloning.

PLANT AND ANIMAL CLONING

Most plants can be cloned quite easily from pieces of root, stem or leaf. Animals cannot be cloned in the same way from parts of their bodies. If animal embryos are divided up at an early stage into several pieces, each piece can develop into a separate animal. (This happens naturally when identical twins are formed.) However, it is hard to predict which embryos will develop into animals with desirable characteristics and should therefore be cloned. The first successful reproductive cloning of an adult with known characteristics produced Dolly the sheep (below).

Techniques for cloning using differentiated cells

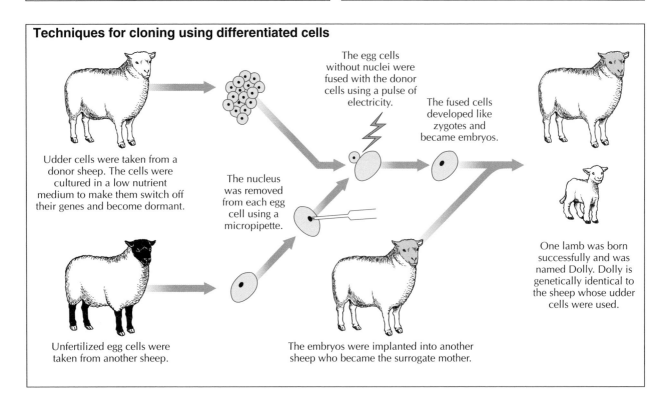

Udder cells were taken from a donor sheep. The cells were cultured in a low nutrient medium to make them switch off their genes and become dormant.

The nucleus was removed from each egg cell using a micropipette.

The egg cells without nuclei were fused with the donor cells using a pulse of electricity.

The fused cells developed like zygotes and became embryos.

Unfertilized egg cells were taken from another sheep.

The embryos were implanted into another sheep who became the surrogate mother.

One lamb was born successfully and was named Dolly. Dolly is genetically identical to the sheep whose udder cells were used.

THERAPEUTIC CLONING IN HUMANS

Techniques are being developed to create human embryos, from which embryonic stem cells can be obtained for medical use. These stem cells have the capacity to divide and differentiate into any types of human cell. They could be used to replace tissues or even organs that have become damaged or lost in a patient. There are many ethical issues involved and research into therapeutic cloning has been banned in some countries.

Arguments for therapeutic cloning

1. Embryonic stem cells can be used for therapies that save lives and reduce suffering.

2. Cells can be removed from embryos that have stopped developing, so would have died anyway.

3. Cells are removed at a stage when embryos have no nerve cells and cannot feel pain.

Arguments against therapeutic cloning

1. Every human embryo is a potential human being, which should be given a chance of developing.

2. More embryos may be produced than are needed, so some may have to be killed.

3. There is a danger of embryonic stem cells developing into tumour cells.

THE HUMAN GENOME PROJECT

The human genome has been estimated to consist of between 25 000 and 30 000 genes. The Human Genome Project aims to find the location of all of these genes on the human chromosomes and the base sequence of all of the DNA that makes them up. The project is an international cooperative one, with laboratories in many countries involved.

The sequencing of the entire human genome will make it easier to study how genes influence human development. It will allow easier identification of genetic diseases. It will allow the production of new drugs based on DNA base sequences of genes or the structure of proteins coded for by these genes. It will give us new insights into the origins, evolution and migrations of humans.

EXAM QUESTIONS ON TOPIC 4

1 In humans the blood groups A, B, AB and O are determined by three alleles of an autosomal gene: I^A, I^B and i. Alleles I^A and I^B are codominant and allele i is recessive. The phenotypes of some individuals in the pedigree below are shown.

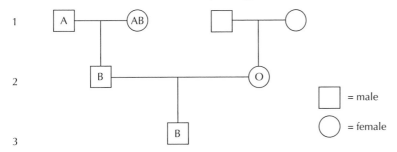

= male

= female

a) Explain the conclusions that can be drawn about the genotypes of the individuals in the pedigree in generations 2 and 3. [3]

b) Explain to which blood groups the parents of the blood group O female in the pedigree could have belonged. [3]

c) Suggest one reason for testing the blood groups in humans. [1]

2 When red and white flowered *Mirabilis jalapa* plants are crossed together, all the offspring have pink flowers. The symbols for the two alleles involved are C^r (red) and C^w (white).

a) State the genotypes of the red- and white-flowered parents and the pink-flowered offspring. [1]

b) When Mendel crossed red- and white-flowered pea plants together, all of the offspring had red flowers. Suggest a reason for the difference in results between pea plants and *Mirabilis jalapa* plants. [1]

c) Predict the outcome of a cross between two pink-flowered *Mirabilis jalapa* plants, using a Punnett grid. [3]

3 a) Define clone. [1]

b) Outline one technique for cloning animals, using differentiated cells. [2]

The DNA profiles of sheep are shown (right).

   U = differentiated cells taken from the udder of a sheep that was used in cloning experiments

   C = cells in a culture derived from the udder cells

   D = blood cells taken from Dolly the sheep

   1–12 = results from 12 other sheep.

c) (i) Explain whether DNA fragments in the profiles had moved upwards or downwards. [2]

   (ii) Explain the conclusions that can be drawn from the DNA profiles of the sheep. [3]

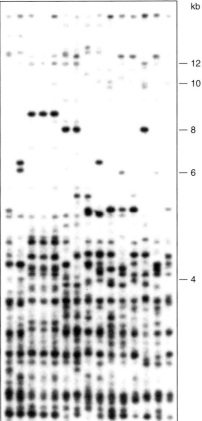

Identifying living organisms

USING KEYS TO IDENTIFY ORGANISMS

The first stage in many ecological investigations is to find out what species of organism there are in the area being studied. This is called **species identification**. This can be done using **keys**.

Keys for species identification are usually constructed in this way:
- the key consists of a series of numbered stages
- each stage consists of a pair of alternative characteristics
- some alternatives give the next stage of the key to go to
- some alternatives give the identification.

Identifying aquarium plants using a key

Many aquatic plants in aquariums in biology laboratories belong to one of these four genera:
- *Cabomba*
- *Ceratophyllum*
- *Elodea*
- *Myriophyllum*

All of these plants have cylindrical stems with whorls of leaves. The shape of four leaves is shown in the figure (left). A key can be used to identify which of the four genera a plant belongs to, if it is known to be in one of them.

1. Simple undivided leaves . . . .*Elodea*

   Leaves forked or divided into segments 2

2. Leaves forked once or twice to form
   two or four segments . . . .*Ceratophyllum*

   Leaves divided into more than four segments 3

3. Leaves divided into many flattened segments . . . .*Cabomba*

   Leaves divided into many
   filamentous segments . . . .*Myriophyllum*

Some species of Elodea have recently been moved by taxonomists to other genera:
Elodea densa is now *Egeria densa.*
Elodea crispa is now *Lagarosiphon major.*

Leaves of aquarium plants

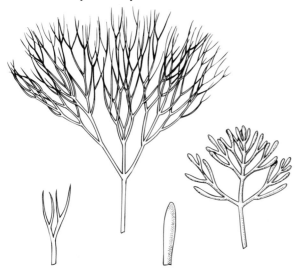

Constructing a key

The five animals shown below are found in beehives. It would be useful to construct a key to allow a beekeeper to identify them, as some of them are very harmful and others are harmless to honey bees.

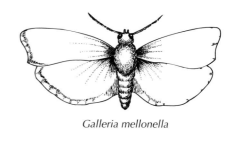

Galleria mellonella

Acarus siro

Braula coeca

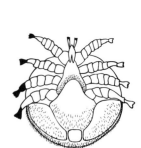

Acarapis woodi

Varroa jacobsonii

BINOMIAL NOMENCLATURE

In the classification of living organisms, the basic group is the **species**.

A species is a group of organisms with similar characteristics, which can interbreed and produce fertile offspring.

Every species is classified into a **genus**. A genus is a group of similar species.

Each species needs an international name, so that biologists throughout the world can refer to it. The naming of species is called **nomenclature**. The nomenclature that biologists use is called the **binomial system** because two names are used to refer to each species. The key features of the binomial system of nomenclature are:
- the first name is the genus name
- the genus name is given an upper case first letter
- the second name is the species name
- the species name is given a lower case first letter
- italics are used when the name is printed
- the name is underlined if it is hand-written.

Classification of plants and animals

CLASSIFICATION FROM SPECIES TO KINGDOM

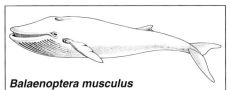

Balaenoptera musculus

A group of organisms, such as a species or a genus is called a **taxon**. Species are classified into a series of taxa, each of which includes a wider range of species than the previous one. This is called the **hierarchy of taxa**.

Sequoia sempervirens

	Animal example **Balaenoptera musculus** **– the blue whale (left)**	**Plant example** **Sequoia sempervirens** **– the coast redwood (right)**
Species that are similar are grouped into a genus	Genus Balaenoptera	Genus Sequoia
Genera that are similar are grouped into a family	Family Balaenopteridae	Family Taxodiaceae
Families that are similar are grouped into an order	Order Cetacea	Order Pinales
Orders that are similar are grouped into a class	Class Mammalia	Class Pinopsida
Classes that are similar are grouped into a phylum	Phylum Chordata	Phylum Coniferophyta
Phyla that are similar are grouped into a kingdom	Kingdom Animalia	Kingdom Plantae

Plant classification

There are four main phyla of plants, which can be easily distinguished by studying their external structure.

	Roots, leaves and stems	**Maximum height**	**Reproductive structures**
Bryophytes – mosses	Bryophytes have no roots, only structures similar to root hairs called **rhizoids** Mosses have simple leaves and stems Liverworts consist of a flattened thallus	0.5 metres	Spores are produced in a capsule. The capsule develops at the end of a stalk
Filicinophytes – ferns	Ferns have roots, leaves and short non-woody stems. The leaves are usually curled up in bud and are often **pinnate** – divided into pairs of leaflets	15 metres	Spores are produced in sporangia, usually on the underside of the leaves
Coniferophytes – conifers	Conifers are shrubs or trees with roots, leaves and woody stems. The leaves are often narrow with a thick waxy cuticle	100 metres	Seeds are produced. The seeds develop from ovules on the surface of the scales of female **cones**. Male cones produce pollen
Angiospermophytes – flowering plants	Flowering plants are very variable but usually have roots, leaves and stems. The stems of flowering plants that develop into shrubs and trees are woody	100 metres	Seeds are produced. The seeds develop from ovules inside **ovaries**. The ovaries are part of **flowers**. **Fruits** develop from the ovaries, to disperse the seed

ANIMAL CLASSIFICATION

There are over thirty phyla of animals. The external recognition features of six of these phyla are shown here.

Porifera
- no clear symmetry
- attached to a surface
- pores through body
- no mouth or anus
Examples: *sponges*

Cnidaria
- radially symmetric
- tentacles
- stinging cells
- mouth but no anus
Examples: *jellyfish, corals, sea anemones*

Platyhelminths
- bilaterally symmetric
- flat bodies
- unsegmented
- mouth but no anus
Examples: *Planaria, tapeworms, liverflukes*

Annelida
- bilaterally symmetric
- bristles often present
- segmented
- mouth and anus
Examples: *earthworms, leeches, ragworms*

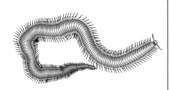

Mollusca
- muscular foot and mantle
- shell usually present
- segmentation not visible
- mouth and anus
Examples: *slugs, snails, clams, squids*

Arthropoda
- bilaterally symmetric
- exoskeleton
- segmented
- jointed appendages
Examples: *insects, spiders, crabs, millipedes*

Population dynamics

CHANGES TO THE SIZE OF A POPULATION

A population is a group of organisms of the same species, who live in the same area at the same time.

There are four ways in which the size of a population can change:
- Offspring are produced and are added to the population – **natality**.
- Individuals die and are lost from the population – **mortality**.
- Individuals move into the area from elsewhere and are added to the population – **immigration**.
- Individuals move out of the area to live elsewhere – **emigration**.

Populations are often affected by all four of these things and the overall change can be calculated using an equation:

$$\text{Population change} = (\text{natality} + \text{immigration}) - (\text{mortality} + \text{emigration})$$

POPULATION GROWTH CURVES

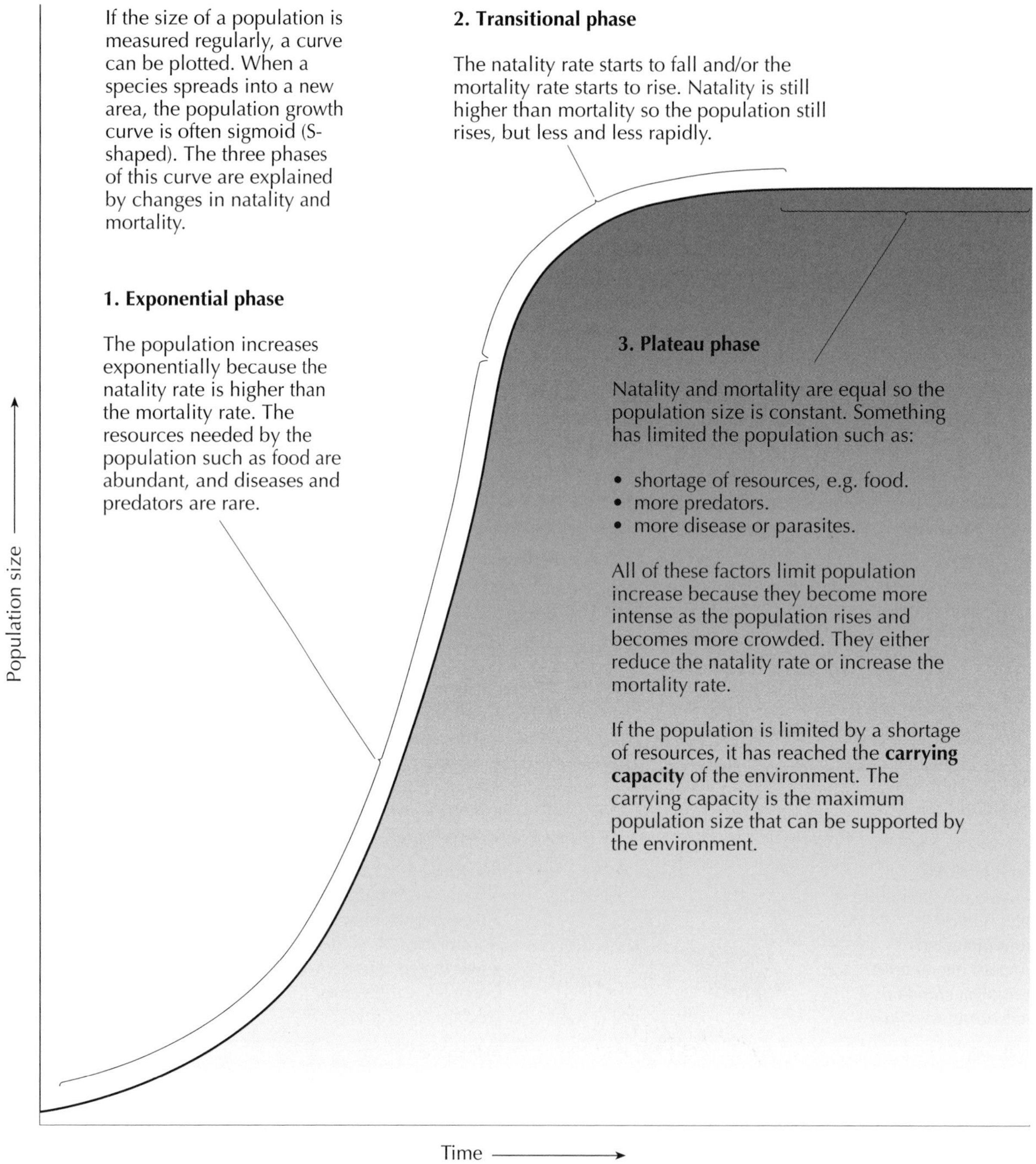

If the size of a population is measured regularly, a curve can be plotted. When a species spreads into a new area, the population growth curve is often sigmoid (S-shaped). The three phases of this curve are explained by changes in natality and mortality.

1. Exponential phase

The population increases exponentially because the natality rate is higher than the mortality rate. The resources needed by the population such as food are abundant, and diseases and predators are rare.

2. Transitional phase

The natality rate starts to fall and/or the mortality rate starts to rise. Natality is still higher than mortality so the population still rises, but less and less rapidly.

3. Plateau phase

Natality and mortality are equal so the population size is constant. Something has limited the population such as:

- shortage of resources, e.g. food.
- more predators.
- more disease or parasites.

All of these factors limit population increase because they become more intense as the population rises and becomes more crowded. They either reduce the natality rate or increase the mortality rate.

If the population is limited by a shortage of resources, it has reached the **carrying capacity** of the environment. The carrying capacity is the maximum population size that can be supported by the environment.

Population size (y-axis)

Time ⟶ (x-axis)

Evidence for evolution

EVOLUTION OF POPULATIONS

The word evolution has several meanings, all of which involve the gradual development of things. In biology, the word has come to mean the changes that occur in living organisms, over many generations. Evolution happens in populations of living organisms. It only happens with characteristics that can be inherited.

Evolution is the cumulative change in the heritable characteristics of a population.

Although it is not possible to prove, using the scientific method, that the organisms on Earth today are the result of evolution, there is much evidence that makes it very likely. Three types of evidence are illustrated on this page.

HOMOLOGOUS ANATOMICAL STRUCTURES

There are also remarkable similarities between some groups of organisms in their structure. For example, bones in the limbs of vertebrates are strikingly similar, despite being used in many different ways (below). The structure is called the pentadactyl limb.

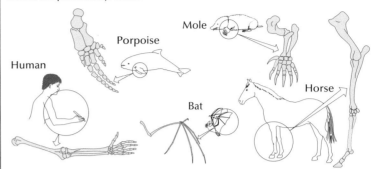

The most likely explanation for these structural similarities is that the organisms have evolved from a common ancestor. Structures that have developed from the same part of a common ancestor are called homologous structures.

THE FOSSIL RECORD – PALAEONTOLOGY

Fossil of *Acanthostega*

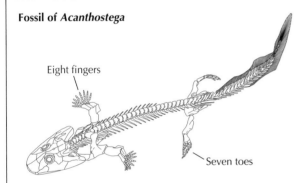

Eight fingers

Seven toes

The existence of fossils is very difficult to explain without evolution. An example of this is *Acanthostega*. The figure (left) is a drawing of a 365 million year old fossil of *Acanthostega*. It has similarities to other vertebrates, with a backbone and four limbs, but it has eight fingers and seven toes, so it is not identical to any existing organism. This suggests that vertebrates and other organisms change over time. *Acanthostega* is an example of a "missing link". Although it has four legs, like most amphibians, reptiles and mammals, it also had a fish-like tail and gills and lived in water. This shows that land vertebrates could have evolved from fish via an aquatic animal with legs.

SELECTIVE BREEDING OF DOMESTICATED ANIMALS

The breeds of animal that are reared for human use are clearly related to wild species and in many cases can still interbreed with them. These domesticated breeds have been developed from wild species, by selecting individuals with desirable traits, and breeding from them. The striking differences in the heritable characteristics of domesticated breeds give us evidence that species can evolve rapidly.

Spanish, Hamburgh and Polish Fowl, illustrated in *Breeds of Animals and Plants under Domestication* by Charles Darwin

Natural selection

DARWIN, WALLACE AND EVOLUTION BY NATURAL SELECTION

Charles Darwin developed the theory that evolution occurs as a result of natural selection. He explained his theory in *The Origin of Species*, published in 1859. He had done many years of research and had collected much evidence for the theory before then.

Darwin delayed publication of his ideas for many years, fearing a hostile reaction. He might never have published them if another biologist, Alfred Wallace, had not written a letter to him in 1858 suggesting very similar ideas.

The theory of evolution by natural selection can be explained in a series of observations and deductions.

The photograph on the right shows a statue of Charles Darwin at Shrewsbury School, where he was a pupil from 1818 to 1825.

Observations	Deductions
* Populations of living organisms tend to increase exponentially * Yet, on the whole, the number of individuals in populations remains nearly constant	* More offspring are produced than the environment can support * There is a struggle for existence in which some individuals survive and some die
* Living organisms vary. The members of a species are different from each other in many ways * Some individuals have characteristics that make them well adapted to their environment and other individuals have characteristics that make them less well adapted to their environment	* The better adapted individuals tend to survive and reproduce more than the less well-adapted individuals **This is natural selection**
* Much variation is heritable – it can be passed on to offspring	* The better-adapted individuals pass on their characteristics to more offspring than the less well adapted individuals. The results of natural selection therefore accumulate * As one generation follows another, the characteristics of the species gradually change – the species evolves

In 1828 Darwin, as a young man was struggling to learn enough mathematics to pass a university exam.

The extract below is from a letter that he wrote to Charles Whitley, a friend and eminent mathematician.

' I am as idle as idle can be: one of the causes you have hit on, viz irresolution the other being made fully aware that my noddle is not capacious enough to retain or comprehend Mathematics. – Beetle hunting & such things I grieve to say is my proper sphere…'

Evolution in action

ENVIRONMENTAL CHANGE AND EVOLUTION

Since Darwin developed his theory of evolution by natural selection, changes have been observed in some species. In each case, the evolution has been in response to environmental change.

Two examples are described here – the development of antibiotic resistance in bacteria and melanism in ladybugs. Other examples include the development of metal tolerance in plants growing on waste material from mining metal ores, and changes to the beaks of finches on the Galápagos Islands in response to El Niño events.

All these recent cases of observed evolution involve relatively small changes, but they do nonetheless add to the evidence for evolution.

SEXUAL REPRODUCTION AND EVOLUTION

Variation is essential for natural selection and therefore for evolution. Although mutation is the original source of new genes or alleles, sexual reproduction promotes variation by allowing the formation of new combinations of alleles. Two stages in sexual reproduction promote variation.

1. Meiosis allows a huge variety of genetically different gametes to be produced by each individual.
2. Fertilization allows alleles from two different individuals to be brought together in one new individual.

Prokaryotes do not reproduce sexually but have other ways to promote variation by exchanging genes.

Some species of organisms only reproduce asexually. Mutations still produce some variation in these species, but without sexual reproduction the variation and the capacity for evolution is less.

Multiple antibiotic resistance in bacteria

Antibiotics are used to control diseases caused by bacteria in humans. There have been increasing problems with disease-causing bacteria being resistant to antibiotics. The figure below shows the percentage of cases of gonorrhea (a sexually transmitted disease) in the United States that were caused by antibiotic-resistant strains of *Neisseria gonorrhoeae* between 1980 and 1990. The trend with many other diseases has been similar.

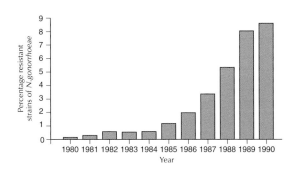

Genes that give resistance to an antibiotic can be found in the micro-organisms that naturally make that antibiotic. The evolution of multiple antibiotic resistance involves the following steps.

- A gene that gives resistance to an antibiotic is transferred to a bacterium by means of a plasmid or in some other way. There is then variation in this type of bacterium – some of the bacteria are resistant to the antibiotic and some are not.
- Doctors or vets use the antibiotic to control bacteria. Natural selection favours the bacteria that are resistant to it and kills the non-resistant ones.
- The antibiotic-resistant bacteria reproduce and spread, replacing the non-resistant ones. Eventually, most of the bacteria are resistant.
- Doctors or vets change to a different antibiotic to control bacteria. Resistance to this soon develops, so another antibiotic is used, and so on until multiply resistant bacteria have evolved.

The more an antibiotic is used, the more bacteria resistant to it there will be and the fewer non-resistant.

Evolution of melanism in ladybugs

Adalia bipunctata, the two-spot ladybug (or ladybird), is a small beetle, which usually has red wing cases with two black spots. The red colour warns predators that it tastes unpleasant. Melanic forms also exist, with black wing cases. The melanic form absorbs heat more efficiently than the red form. It therefore has a selective advantage when sunlight levels are low and it is difficult for ladybugs to warm up. The melanic form of *Adalia bipunctata* became common in industrial areas of Britain, but declined again after 1960. The decline correlates with decreases in smoke in the air (below). In air darkened by smoke, the melanic forms will be able to warm up more quickly, but if the smoke is no longer present this advantage is lost and warning colouration is more important.

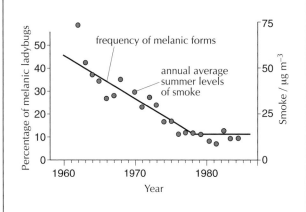

Adalia 2-punctata
(f. *typica*)
2-spot ladybird (typical)

Adalia 2-punctata
(f. *quadrimaculata*)
2-spot ladybird (melanic)

Trophic levels

Populations do not live in isolation – they live together with other populations in **communities**. *A community is a group of populations living together and interacting with each other in an area.*

There are many types of interaction between populations in a community. Trophic relationships are very important – where one population of organisms feeds on another population. Sequences of trophic relationships, where each member in the sequence feeds on the previous one, are called **food chains**.

An example from rainforest at Iguazu in north-east Argentina.

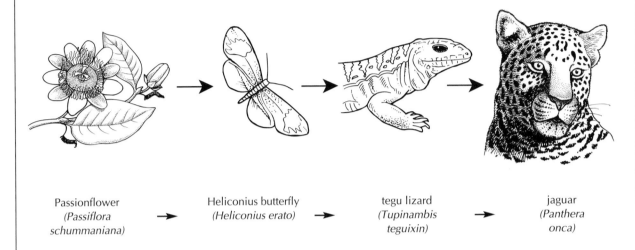

| Passionflower (Passiflora schummaniana) | → | Heliconius butterfly (Heliconius erato) | → | tegu lizard (Tupinambis teguixin) | → | jaguar (Panthera onca) |

An example from chalk grassland and the air above it in Europe.

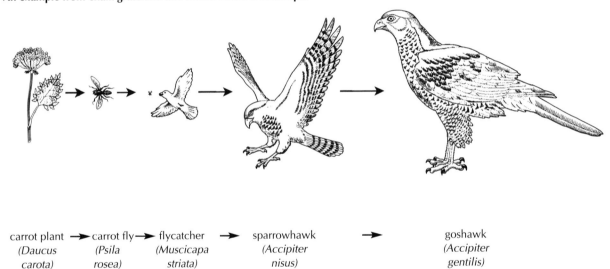

carrot plant → carrot fly → flycatcher → sparrowhawk → goshawk
(Daucus carota) (Psila rosea) (Muscicapa striata) (Accipiter nisus) (Accipiter gentilis)

The first organism in a food chain does not feed on other organisms so must be a producer – an organism that makes its own food. The other organisms are all consumers and are called primary, secondary, tertiary and so on, depending on their position in the chain.

Producer, primary consumer, secondary consumer and tertiary consumer are examples of **trophic levels**. *The trophic level of an organism is its position in the food chain.*

Example:

| **Producer** | → | **Primary consumer** | → | **Secondary consumer** | → | **Tertiary consumer** |
| Sea lettuce (Ulva lactuca) | | Marine iguana (Amblyrhyncus cristatus) | | Galapagos snake (Dromiscus biserialis) | | Galapagos hawk (Buteo galapagensis) |

A food chain shows only some of the trophic relationships in a community. Organisms rarely feed on only one other organism and are usually fed on by more than one organism. The complex network of trophic relationships in a community is shown in full in a complex diagram called a food web. An example of a food web is shown on page 42.

Energy flow

AUTOTROPHS

The organisms in a community all need a supply of energy. Organisms are divided into two groups according to their source – **autotrophs** and **heterotrophs**.

Autotrophs are organisms that synthesize their own organic molecules (food) from simple inorganic substances.

Autotrophs make their own food, so are also called producers. Oak trees, maize plants, algae, blue-green bacteria are examples. All food chains start with a producer. In almost all communities, the producers make organic matter by photosynthesis.

Light is therefore the **initial energy source** for the whole community. Producers convert light energy into the chemical energy of sugars and other organic compounds. This energy trapped by the producers eventually leaves them in one of three ways, shown in the flow chart (right).

Energy flow through producers

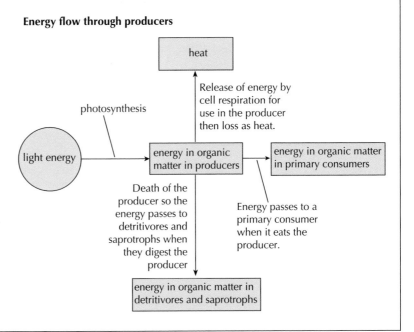

HETEROTROPHS

Heterotrophs are organisms that obtain organic molecules (food) from other organisms.

There are three types of heterotroph: **consumers, detritivores** and **saprotrophs**.

Consumers are organisms that ingest organic matter that is living or recently killed. Primary consumers eat producers and so obtain energy from them. They do not absorb all of the energy in the food that they eat. The energy that they do not take into their tissues leaves them in one of three ways, shown in the flow chart (right). There are similar energy losses from secondary and tertiary consumers in the food chain. Locusts, sheep and lions are examples of consumers.

Detritivores ingest dead organic matter. Dung beetles and earthworms are examples of detritivores.

Saprotrophs live on or in dead organic matter, secreting enzymes into it and absorbing the products of digestion. Bread mould and mushrooms are examples of saprotrophs.

The energy that passes to detritivores and saprotrophs is eventually released by cell respiration and lost as heat. In most communities all the light energy that was trapped by producers is ultimately lost as heat after flowing through the food chain. A summary of energy flow for a three-stage food chain is shown (right).

Energy flow through consumers

Energy flow through a food chain

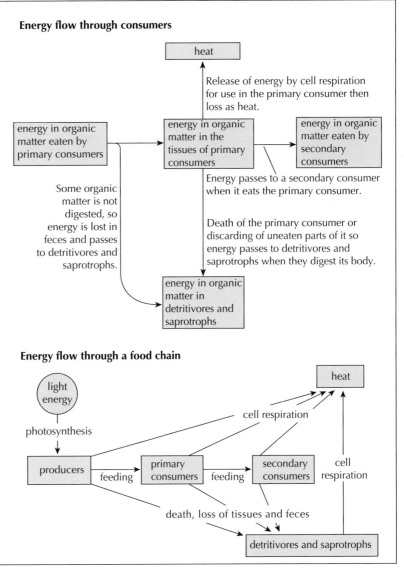

Food webs and energy pyramids

FOOD WEBS

A food web is a diagram that shows all the feeding relationships in a community. The arrows indicate the direction of energy flow. Complete food web diagrams are very complex. The figure (below) shows a simplified food web for a community that lives in an area of Arctic tundra in Ogotoruk Valley.

Food web for Arctic tundra

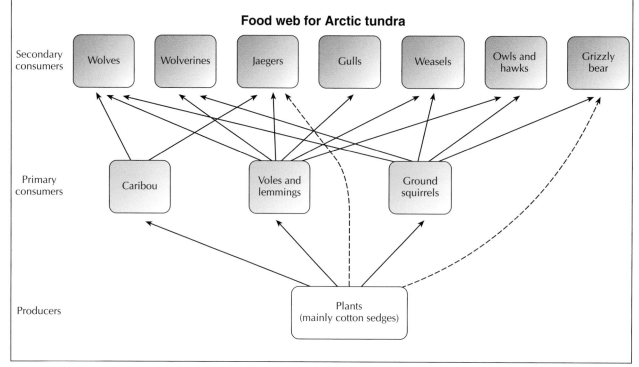

ENERGY PYRAMIDS

Energy pyramids are diagrams that show how much energy flows through each trophic level in a community. The amounts of energy are shown per square metre of area occupied by the community and per year ($kJ\,m^{-2}\,year^{-1}$). The figure (right) is a pyramid of energy for Silver Springs, a stream in Florida.

The figure (below right) is a pyramid of energy for a salt marsh in Georgia. Pyramids of energy are always pyramid shaped – each level is smaller than the one below it. This is because less energy flows through each successive trophic level. Energy is lost at each trophic level, so less remains for the next level. Note that mass is lost as well as energy, so the energy content per gram of the tissues of each successive trophic level is *not* lower.

Energy is lost in various ways. In each of the first three ways the energy is not completely lost from the community as it passes to detritivores and saprotrophs.

- Some organisms die before an organism in the next trophic level eats them.
- Some parts of organisms such as bones or hair are not eaten.
- Some parts of organisms are indigestible and pass out as feces.
- Much of the energy absorbed by an organism is released in cell respiration. The energy, in the form of ATP, is used in processes such as muscle contraction or active transport that require energy. These processes involve energy transformations, which are never 100% efficient. Some of the energy is converted to heat. 10–20% is a typical efficiency level. Most of the energy released by cell respiration is lost from the organism as heat.

Energy absorbed by living organisms is only available to the next trophic level if it remains as chemical energy in the growth of the organism. This is only a small proportion of the energy absorbed.

Energy pyramid for a stream

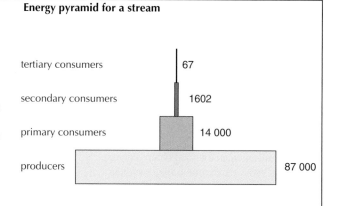

Energy pyramid for a salt marsh

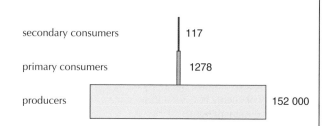

Nutrient recycling

ECOSYSTEMS, ECOLOGISTS AND ECOLOGY

Communities of living organisms interact in many ways with the soil, water and air that surround them. The non-living surroundings of a community are its abiotic environment.
A community and its **abiotic environment** function together as a system called an **ecosystem**. *An ecosystem is a community and its abiotic environment.*
Ecologists study the complex relationships within ecosystems. This area of study is called **ecology**. *Ecology is the study of relationships in ecosystems – both relationships between organisms and between organisms and their environment.*

NUTRIENT RECYCLING IN ECOSYSTEMS

The recycling of nutrients is one example of the interactions between living organisms and the abiotic environment in an ecosystem. Energy is not recycled. It is supplied to ecosystems in the form of light, flows through food chains and is lost as heat. Nutrients are not usually resupplied to ecosystems – they must be used again and again by recycling. Carbon, nitrogen, phosphorus and all the other essential elements must be recycled. They are absorbed from the environment, used by living organisms and then returned to the environment.
The processes involved in the carbon cycle are shown below.

The carbon cycle

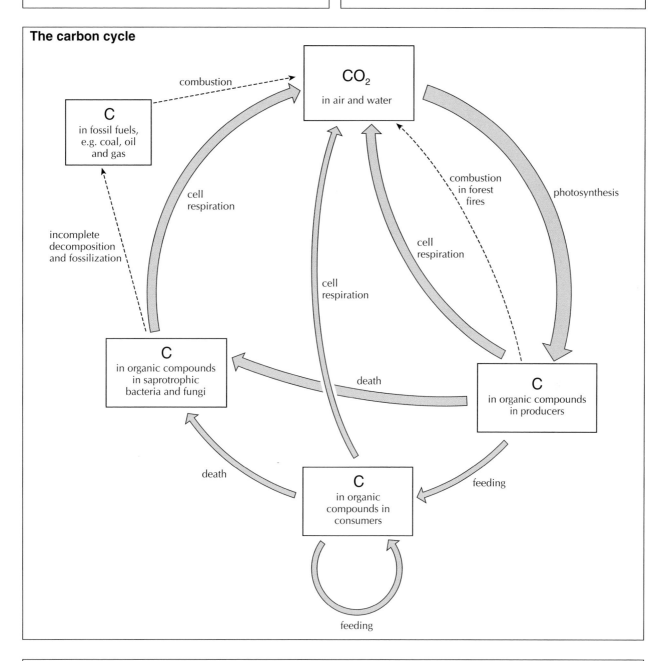

THE ROLE OF SAPROTROPHS IN RECYCLING OF NUTRIENTS

Saprotrophic bacteria and fungi have an essential role in nutrient cycles. They feed by secreting digestive enzymes into dead organic matter, including dead plants and animals and feces. The enzymes gradually break down the organic matter and the nutrients that were locked up in complex organic compounds are released. The saprotrophs absorb the substances that they need from the digested organic matter.
Without saprotrophs, nutrients would remain locked up permanently in dead organic matter and organisms that need the nutrients would soon become deficient.

Global warming and the greenhouse effect

RISING CARBON DIOXIDE LEVELS

The carbon dioxide concentration of bubbles of air trapped in Antarctic ice at different dates have been measured. These show that for two thousand years before 1880 the carbon dioxide concentration of the atmosphere remained fairly constant at about 270 parts per million (ppm).

From 1880 onwards the concentration rose. Since 1958, the concentration has been monitored continuously at Mauna Loa, Hawaii (right). There is an annual fluctuation, but the overall trend has been upwards and the concentration is now more than 100ppm higher than in 1880.

GREENHOUSE GASES

Carbon dioxide is one of a group of gases that cause heat to be retained in the Earth's atmosphere:
• carbon dioxide
• methane
• oxides of nitrogen (NOX)
• sulfur dioxide

Heat retention by gases is called the **greenhouse effect**. The processes involved are shown in the figure (right).

The greenhouse effect is not a new phenomenon. Natural processes produce greenhouse gases, so they are a natural part of the Earth's atmosphere. The change in recent years is that human activities have increased the production of greenhouse gases and so their atmospheric concentrations and their contributions to the greenhouse effect have been rising (below right). This is correlated with rising temperatures on Earth – **global warming**.

RISING GLOBAL TEMPERATURES

Temperature records have been analysed to find the mean for the whole world in each year from 1850 onwards. The figure (right) shows the difference between the mean temperature for each year and an overall mean temperature for the years 1961–1990.

The trends are that
• From 1856 until about 1910 temperatures were relatively stable.
• From 1910 until 1940 temperatures rose and were then stable.
• From 1970 there has been a rapid rise.
• All ten of the hottest years since records began in 1850 have been since 1990.
• 1998 was the hottest year in that period and 2005 was the second hottest year.
• Over the past century, global temperatures have risen by 0.7°C on average, which takes us out of the range of average temperatures experienced on Earth over the last 1000 years.

These changes in temperature are statistically significant. There could be various causes, but the most likely cause is an increased greenhouse effect, due to human activities.

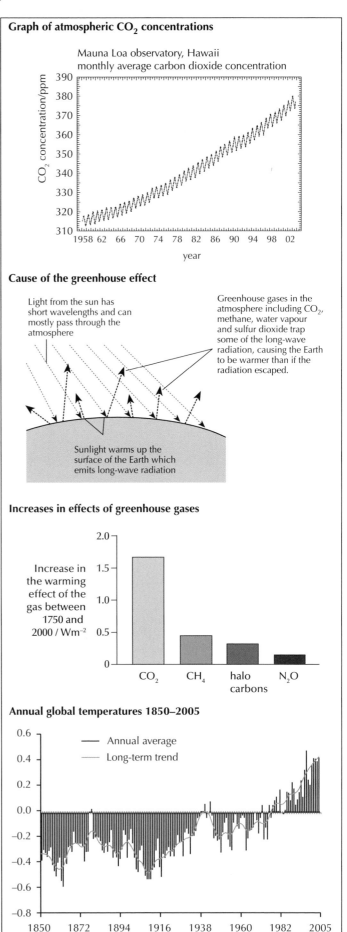

Graph of atmospheric CO_2 concentrations

Mauna Loa observatory, Hawaii monthly average carbon dioxide concentration

Cause of the greenhouse effect

Light from the sun has short wavelengths and can mostly pass through the atmosphere

Greenhouse gases in the atmosphere including CO_2, methane, water vapour and sulfur dioxide trap some of the long-wave radiation, causing the Earth to be warmer than if the radiation escaped.

Sunlight warms up the surface of the Earth which emits long-wave radiation

Increases in effects of greenhouse gases

Increase in the warming effect of the gas between 1750 and 2000 / Wm^{-2}

CO_2 CH_4 halo carbons N_2O

Annual global temperatures 1850–2005

— Annual average
······· Long-term trend

Responses to global warming

HABITATS

The Earth provides places for millions of living organisms to exist. These places are called **habitats**.

A habitat is the environment in which a species normally lives or the location of a living organism.

CONSEQUENCES OF GLOBAL WARMING

The effects of global warming are already being felt, but they are likely to become much more extreme during the 21st century. Habitats throughout the world will be affected, but the effects on Arctic ecosystems could be particularly catastrophic.

- Glaciers will melt and polar ice sheets will break up into icebergs, which will also eventually melt. The Arctic ice cap may disappear completely.
- Permafrost will melt during the summer, increasing the rates of decomposition of trapped organic matter, including peat and detritus. This will cause release of carbon dioxide, further increasing atmospheric concentrations.
- Species adapted to temperate conditions will spread north, altering food chains and affecting animals in the higher trophic levels.
- Marine species of animal in Arctic waters may become extinct, as they can be very sensitive to temperature changes in seawater.
- Polar bears and other animals will lose their ice habitat, where they feed and breed.
- Pests and diseases may become more prevalent, with warmer temperatures.
- Sea levels will rise and low-lying areas of land will be flooded.
- Extreme weather events, such as storms, will become more frequent, with harmful effects on species that are not adapted.

THE PRECAUTIONARY PRINCIPLE

In a court of law, prosecutors try to prove that the defendant is guilty. If they cannot do this, the defendant is assumed to be innocent. When the precautionary principle is followed, the opposite policy is adopted – people planning to do something must prove that it will *not* do harm, before actually doing it. The precautionary principle should be followed when the possible consequences (risks) of human actions are very large or could even be catastrophic.

Although there is strong evidence that greenhouse gas emissions are causing global warming, there is no proof. Some politicians and business leaders have argued against measures to combat global warming, because it is not *certain* that greenhouse gases are causing it. Oil companies and airlines in particular have voiced opposition.

Many scientists have argued that if we waited for proof of the effects of greenhouse gas emissions before reacting, the consequences would probably have reached a catastrophic level. The risks are so great that the precautionary principle should be followed: anyone advocating continuing to emit greenhouse gases at current levels, or even to increase emissions, should be required to prove that this will not cause a damaging increase in the greenhouse effect.

Ice sheet breaking up, with Greenland in the background

Musk ox (*Ovibos moschatus*) a species of mammal which nearly became extinct through over-hunting in the early 20th century. Protection by the Canadian government has allowed them to increase, with the population now over 60 000. They have a very thick coat and are adapted to the cold conditions of Arctic regions. The consequence of global warming for musk ox can be deduced from the distribution map (below).

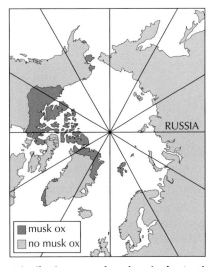

RUSSIA

- musk ox
- no musk ox

Distribution map of musk ox in the Arctic

1 The graph below shows the growth of a population of ring-necked pheasants (*Phasianus colchicus*) on Protection Island off the north west coast of the United States. The original population released by the scientists consisted of two male and eight female birds. Two of the females died immediately after release.

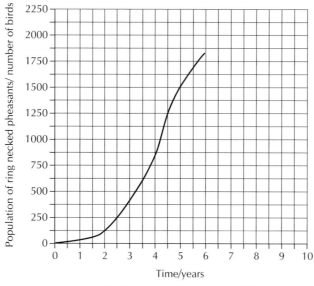

[Source of data: Elinarson A. S., Murrelet, (1945) 26: pages 39–44]

a) State the term used to describe the shape of a growth curve of this type. [1]

b) (i) The scientists predicted that the population would reach its carrying capacity of 2000 by year 8.
Draw a line on the graph to show the population growth between years 6 and 10. [2]

c) (i) Predict how the population growth would change if all the female birds in the original sample had survived. [1]

   (ii) Predict the effect on the carrying capacity if all the female birds in the original sample had survived. [1]

2 The diagram below shows in simplified form the transfers of energy in a generalized ecosystem.
Each box represents a category of organisms, grouped together by their trophic position in the ecosystem.

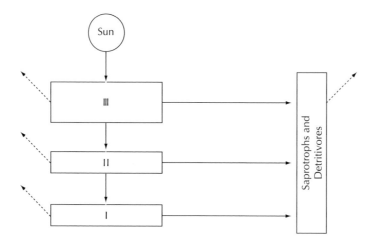

a) Deduce the trophic levels of the organisms in boxes I, II and III. [3]

b) State the form in which energy enters organisms in box 1. [1]

c) Identify which arrow represents the greatest transfer of energy per unit of time. (Add a large X to the arrow). [1]

d) Explain what is represented by the dotted arrows leaving each box. [3]

3 Methane acts as a greenhouse gas in the atmosphere. The main sources of methane are the digestive systems of cattle and sheep, bacterial action in rice paddies, burning of biomass (e.g. forest fires), bacterial action in swamps and marshes, burning of coal and release of natural gas.

a) Discuss whether methane emissions from these sources will cause a change in the Earth's temperature. [3]

b) Discuss whether release of methane is a natural process or an example of a human impact on the environment. [3]

c) Suggest measures that could be taken to reduce the emission of methane. [3]

Digestion

TAKING IN FOOD

Humans take food into their digestive system through the mouth and the esophagus. However, this food is not truly inside the body until it has passed through a layer of cells into the body's tissues. This happens in the small intestines and is called **absorption**. Small finger-like projections from the wall of the small intestine called villi are specially adapted to absorb food molecules. The structure of a villus is shown below. After food has been absorbed it is **assimilated** – it becomes part of the tissues of the body.

Structure of a villus

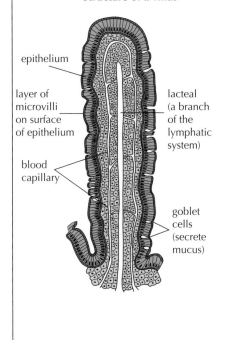

epithelium

layer of microvilli on surface of epithelium

lacteal (a branch of the lymphatic system)

blood capillary

goblet cells (secrete mucus)

RELATIONSHIP BETWEEN STRUCTURE OF A VILLUS AND ITS FUNCTION

- Villi increase the surface area over which food is absorbed.
- An epithelium, consisting of only one thin layer of cells, is all that foods have to pass through to be absorbed.
- Protrusions of the exposed part of the plasma membranes of the epithelium cells increase the surface area for absorption. These projections are called microvilli.
- Protein channels in the microvilli membranes allow rapid absorption of foods by facilitated diffusion and pumps allow rapid absorption by active transport.
- Mitochondria in epithelium cells provide the ATP needed for active transport.
- Blood capillaries inside the villus are very close to the epithelium so the distance for diffusion of foods is very small.
- A lacteal (a branch of the lymphatic system) in the centre of the villus carries away fats after absorption.

THE NEED FOR DIGESTION

The food that humans eat contains substances made by other organisms, many of which are not suitable for human tissues. They must therefore be broken down and reassembled in a form that is suitable.

A second reason for digestion is that many of the molecules in foods are too large to be absorbed by the villi in the small intestine. These large molecules have to be broken down into small molecules that can then be absorbed by diffusion, facilitated diffusion or active transport. The three main types of food molecule that need to be digested are starch, protein and triglycerides (fats and oils).

Digestion of these large molecules happens naturally at body temperature, but only at a very slow rate. Enzymes are essential to speed up the process.

Enzymes of digestion

	Amylase	Protease	Lipase
Example of this enzyme	Salivary amylase	Pepsin	Pancreatic lipase
Source	Salivary glands	Wall of stomach	Pancreas
Substrate	Starch	Proteins	Triglycerides (fats or oils)
Products	Maltose	Small polypeptides	Fatty Acids and Glycerol
Optimum pH	pH 7	pH 1.5	pH 7

The human digestive system

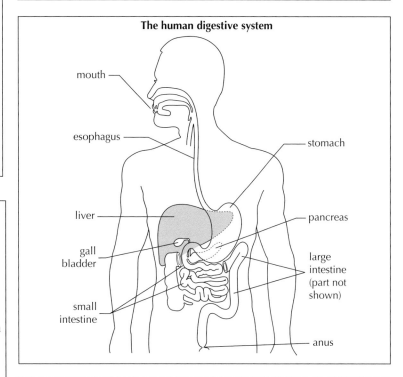

mouth

esophagus

stomach

liver

pancreas

gall bladder

large intestine (part not shown)

small intestine

anus

FUNCTIONS OF THE STOMACH AND INTESTINES

Digestion of proteins begins in the stomach, catalysed by pepsin. Bacteria, which could cause food poisoning, are mostly killed by the acid conditions of the stomach. The acidity also provides optimum conditions for pepsin to work.

Enzymes secreted by the wall of the small intestine complete the process of digestion. The end products of digestion are absorbed by the villi protruding from the wall of the small intestine.

The indigestible parts of the food, together with a large volume of water, pass on into the large intestine. Water is absorbed here leaving solid feces, which are eventually egested through the anus.

The cardiovascular system

HEART STRUCTURE

The heart is a double pump, with the right side pumping blood to the lungs and the left side pumping blood to all other organs. The walls of the heart are composed of cardiac muscle. Contraction of cardiac muscle is **myogenic** – it can contract on its own, without being stimulated by a nerve. There are many capillaries in the muscular wall of the heart. The blood running through these capillaries is supplied by the coronary arteries, which branch off the aorta, close to the semilunar valve. The blood brought by the coronary arteries brings nutrients. It also brings oxygen for aerobic cell respiration, which provides the energy needed for cardiac muscle contraction.

Structure of the heart

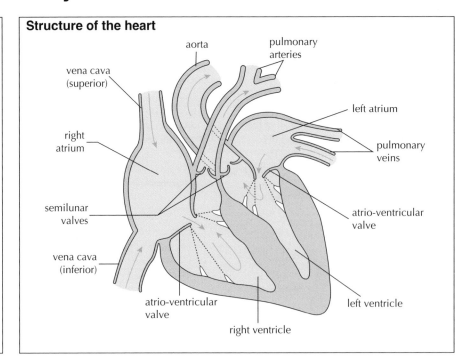

Labels: aorta; pulmonary arteries; vena cava (superior); left atrium; right atrium; pulmonary veins; semilunar valves; atrio-ventricular valve; vena cava (inferior); atrio-ventricular valve; left ventricle; right ventricle

BLOOD VESSELS

Arteries

Thick outer layer of longitudinal collagen and elastic fibres to avoid bulges and leaks

Thick wall to withstand the high pressures

Thick layers of circular elastic and muscle fibres to help pump the blood on after each heart beat

Narrow lumen to help maintain the high pressures

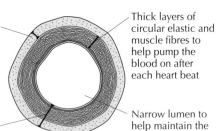

Veins

Thin layers with a few circular elastic and muscle fibres because blood does not flow in pulses so the veins wall cannot help pump it.

Wide lumen is needed to accomodate the slow-flowing blood

Thin wall allows the vein to be pressed flat by adjacent muscles, helping to move the blood

Thin outer layer of longitudinal collagen and elastic fibres because there is little danger of bursting

N.B. veins have valves to prevent back-flow

Capillaries

Wall consists of a single layer of thin cells so the distance for diffusion in or out is small.

Pores between cells in the wall allow some of the plasma to leak out and form tissue fluid. Phagocytes can also squeeze out.

Very narrow lumen – only about 10µm across so that capillaries fit into small spaces. Many small capillaries have a larger surface area than fewer wider ones

THE ACTION OF THE HEART

The atria are the collecting chambers – they collect blood from the veins. The ventricles are the pumping chambers – they pump blood out into the arteries at high pressure. The valves ensure that the blood always flows in the correct direction. Every heartbeat consists of a sequence of actions.

1. The walls of the atria contract, pushing blood from the atria into the ventricles through the atrioventricular valves, which are open. The semilunar valves are closed, so the ventricles fill with blood.
2. The walls of the ventricles contract powerfully and the blood pressure rapidly rises inside them. This rise in pressure first causes the atrioventricular valves to close, preventing back-flow of blood to the atria and then causes the semilunar valves to open, allowing blood to be pumped out into the arteries. At the same time the atria start to refill as they collect blood from the veins.
3. The ventricles stop contracting and as pressure falls inside them the semilunar valves close, preventing back-flow of blood from the arteries to the ventricles. When the ventricular pressure drops below the atrial pressure, the atrioventricular valves open. Blood entering the atrium from the veins then flows on to start filling the ventricles. The next heartbeat begins when the walls of the atria contract again.

THE CONTROL OF THE HEART BEAT

Heart muscle tissue has a special property – it can contract on its own without being stimulated by a nerve. One region is responsible for initiating each contraction. This region is called the pacemaker and is located in the wall of the right atrium. Each time the **pacemaker** sends out a signal the heart carries out a contraction or beat. Nerves and hormones can transmit messages to the pacemaker.

- One nerve carries messages from the brain to the pacemaker that tell the pacemaker to speed up the beating of the heart.
- Another nerve carries messages from the brain to the pacemaker that tell the pacemaker to slow down the beating.
- Adrenalin, carried to the pacemaker by the bloodstream, tells the pacemaker to speed up the beating of the heart.

Blood, transport and infections

THE COMPOSITION OF BLOOD

Blood is composed of plasma, erythrocytes (red blood cells), leukocytes and platelets. The figure below shows the appearance of blood as seen using a light microscope. Two types of leukocyte are shown.

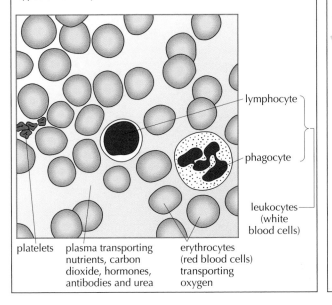

- lymphocyte
- phagocyte
- leukocytes (white blood cells)

platelets | plasma transporting nutrients, carbon dioxide, hormones, antibodies and urea | erythrocytes (red blood cells) transporting oxygen

PHAGOCYTES

Some of the leukocytes in blood are phagocytes. These cells can identify pathogens and ingest them by endocytosis. *A pathogen is an organism or virus that causes disease.* The pathogens are then killed and digested inside the cell by enzymes from lysosomes. Phagocytes can ingest pathogens in the blood. They can also squeeze out through the walls of blood capillaries and move through tissues to sites of infection. They then ingest the pathogens causing the infection. Large numbers of phagocytes at a site of infection form pus.

Some pathogens are able to avoid being killed by phagocytes, so another defence is needed.

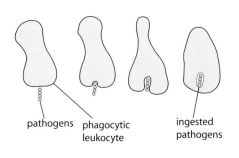

pathogens | phagocytic leukocyte | ingested pathogens

FUNCTIONS OF BLOOD

Blood has two main functions: transport and defence against infectious disease.

Red blood cells transport oxygen from the lungs to respiring cells.

Blood plasma transports
- nutrients
- carbon dioxide
- hormones
- antibodies
- urea.

The blood also transports heat from parts of the body that produce it, to the skin, where it is lost to the environment.

Leukocytes (white blood cells) defend the body against infectious diseases. The roles of phagocytes and leukocytes are described on this page and the next page.

ANTIBODIES

Antibodies are proteins that recognize and bind to specific antigens. Antigens are foreign substances that stimulate the production of antibodies. Antibodies usually only bind to one specific antigen. Antigens can be any of a wide range of substances including cell walls of pathogenic bacteria or fungi and protein coats of pathogenic viruses.

Antibodies defend the body against pathogens by binding to antigens on surface of a pathogen and stimulating its destruction. The figure (on page 50) shows how antibodies are produced.

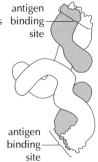

antigen binding site

antigen binding site

Antibody molecule made up of four polypeptides

BARRIERS TO INFECTION

The skin and mucous membranes form a barrier that prevents most pathogens from entering the body. The outer layers of the skin are tough and form a physical barrier. Sebaceous glands in the skin secrete lactic acid and fatty acids, which make the surface of the skin acidic. This prevents the growth of most pathogenic bacteria.

Mucous membranes are soft areas of skin that are kept moist with mucus. Mucous membranes are found in the nose, trachea, vagina and urethra. Although they do not form a strong physical barrier, many bacteria are killed by lysozyme, an enzyme in the mucus. In the trachea pathogens tend to get caught in the sticky mucus and cilia then push the mucus and bacteria up and out of the trachea.

Despite these barriers to infection, pathogens do sometimes enter the body so another defence is needed.

ANTIBIOTICS

Antibiotics are chemicals produced by microorganisms, to kill or control the growth of other microorganisms. For example, *Penicillium* fungus produces penicillin to kill bacteria.

Most bacterial diseases in humans can be treated successfully with antibiotics. For example, tuberculosis has been treated with streptomycin. There are many differences between human cells and bacterial cells and so there are many antibiotics that block a process in bacterial cells without causing any harm to human cells.

Viruses carry out very few processes themselves. They rely instead on a host cell such as a human cell to carry out the processes for them. It is not possible to block these processes with an antibiotic without also harming the human cells. For this reason virus diseases cannot be treated with antibiotics.

Antibodies and AIDS

PRODUCTION OF ANTIBODIES

① Antibodies are made by lymphocytes, one of the two main types of leukocyte.

② A lymphocyte can only make one type of antibody so a huge number of different lymphocyte types is needed. Each lymphocyte puts some of the antibody that it can make into its plasma membrane with the antigen-combining site projecting outwards.

③ When a pathogen enters the body, its antigens bind to the antibodies in the plasma membrane of one type of lymphocyte.

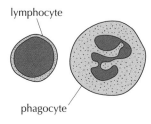

lymphocyte

phagocyte

Variety of antibodies on lymphocyte surfaces.

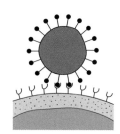

④ When antigens bind to the antibodies on the surface of a lymphocyte, this lymphocyte becomes active and divides by mitosis to produce a clone of many identical cells.

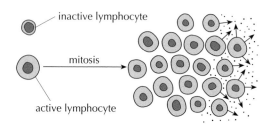

inactive lymphocyte

mitosis

active lymphocyte

⑤ The clone of cells starts to produce large quantities of the same antibody – the antibody needed to defend the body against the pathogen.

AIDS – A SYNDROME CAUSED BY A VIRUS

AIDS shows how vital the body's defences against disease are. Destruction of the immune system leads inevitably to death. AIDS is an example of a syndrome. A syndrome is a group of symptoms that are found together. Individuals with acquired immunodeficiency syndrome (AIDS) have low numbers of one type of lymphocyte together with weight loss and a variety of diseases caused by viruses, bacteria, fungi and protozoa. These diseases weaken the body and eventually cause death.

Cause

HIV (human immunodeficiency virus) causes AIDS. The virus infects a type of lymphocyte that plays a vital role in antibody production. Over a period of years these lymphocytes are destroyed and antibodies cannot then be produced. Without a functioning immune system, the body is vulnerable to pathogens that would normally be controlled easily.

Transmission

HIV does not survive for long outside the body and cannot easily pass through the skin. Transmission involves the transfer of body fluids from an infected person to an uninfected one.

- Through small cuts or tears in the vagina, penis, mouth or intestine during vaginal, anal or oral sex.
- In traces of blood on a hypodermic needle that is shared by intravenous drug abusers.
- Across the placenta from a mother to a baby, or through cuts during childbirth or in milk during breast-feeding.
- In transfused blood or with blood products such as Factor VIII used to treat hemophiliacs.

Social implications

- Families and friends suffer grief.
- Families become poorer if the individual with AIDS was the wage earner and is refused life insurance.
- Individuals infected with HIV may become stigmatized and not find partners, housing or employment.
- Sexual activity in a population may be reduced because of the fear of AIDS.

Structure of HIV

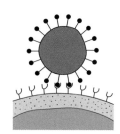

single stranded RNA

reverse transcriptase

lipid bilayer } envelope

glycoprotein

capsid (protein coat)

T-lymphocyte infected with HIV(× 3500)

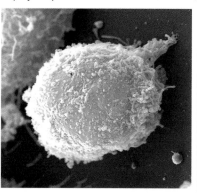

Gas exchange

THE NEED FOR GAS EXCHANGE AND VENTILATION IN HUMANS

Cell respiration happens in the cytoplasm and mitochondria of cells and releases energy in the form of ATP for use inside the cell. In humans oxygen is used in cell respiration and carbon dioxide is produced. Humans therefore must take in oxygen from their surroundings and release carbon dioxide. This process of swapping one gas for another is called **gas exchange**.

Gas exchange happens in the alveoli of human lungs. Oxygen diffuses from the air in the alveoli to the blood in capillaries. Carbon dioxide diffuses in the opposite direction. The figure (below) shows the adaptations of the alveolus for gas exchange. Diffusion of oxygen and carbon dioxide happens because there are concentration gradients of oxygen and carbon dioxide between the air and the blood. To maintain these concentration gradients, the air in the alveoli must be refreshed frequently. The process of bringing fresh air to the alveoli and removing stale air is called **ventilation**.

THE VENTILATION SYSTEM

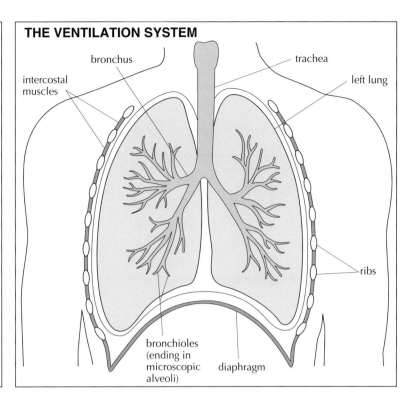

bronchus · trachea · intercostal muscles · left lung · ribs · bronchioles (ending in microscopic alveoli) · diaphragm

ADAPTATIONS OF THE ALVEOLUS TO GAS EXCHANGE

Although each alveolus is very small, the lungs contain hundreds of millions of alveoli in total, giving a huge overall surface area for gas exchange.

The wall of the alveolus consists of a single layer of very thin cells. The capillary wall also is a single layer of very thin cells, so the gases only have to diffuse a very short distance.

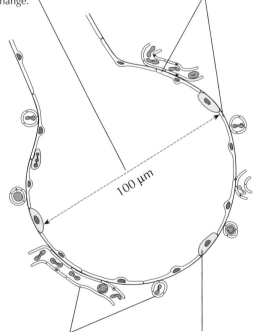

100 μm

The alveolus is covered by a dense network of blood capillaries with low oxygen and high carbon dioxide concentrations. Oxygen therefore diffuses into the blood and carbon dioxide diffuses out.

Cells in the alveolus wall secrete a fluid which keeps the inner surface of the alveolus moist, allowing the gases to dissolve. The fluid also contains a natural detergent, which prevents the sides of the alveoli from sticking together.

VENTILATION OF THE LUNGS

Air is inhaled into the lungs through the trachea, bronchi and bronchioles.

It is exhaled via the same route. Muscles are used to lower and raise the pressure inside the lungs to cause the movements of air.

Inhaling

- The external intercostal muscles contract, moving the ribcage up and out

- The diaphragm contracts, becoming flatter and moving down

- These muscle movements increase the volume of the thorax

- The pressure inside the thorax therefore drops below atmospheric pressure

- Air flows into the lungs from outside the body until the pressure inside the lungs rises to atmospheric pressure

Exhaling

- The internal intercostal muscles contract, moving the ribcage down and in

- The abdominal muscles contract, pushing the diaphragm up into a dome shape

- These muscle movements decrease the volume of the thorax

- The pressure inside the thorax therefore rises above atmospheric pressure

- Air flows out from the lungs to outside the body until the pressure inside the lungs falls to atmospheric pressure

Neurons and synapses

ORGANIZATION OF THE NERVOUS SYSTEM

The nervous system is composed of cells called neurons. These cells are often very elongated and can carry messages at high speed in the form of electrical impulses.
There are two parts of the nervous system
- the central nervous system (CNS), consisting of the brain and spinal cord
- peripheral nerves that connect all parts of the body to the central nervous system.

SENSORY AND MOTOR NEURONS

Neurons carry electrical impulses long distances in the body, using elongated structures called nerve fibres (axons).
- Sensory neurons carry nerve impulses from receptors (sensory cells) to the CNS.
- Motor neurons (below) carry impulses from the CNS to effectors (muscle and gland cells).
- Relay neurons carry impulses within the CNS, from one neuron to another.

Structure of a motor neuron

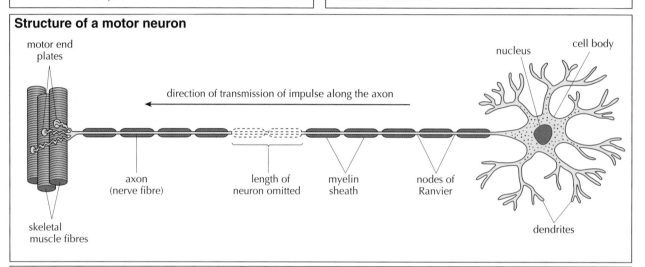

SYNAPSES

A synapse is a junction between two neurons. The plasma membranes of the neurons are separated by a narrow fluid-filled gap called the synaptic cleft. Messages are passed across the synapse in the form of chemicals called **neurotransmitters**. The neurotransmitters always pass in the same direction from the pre-synaptic neuron to the post-synaptic neuron.

Many synapses function in the following way.

1. A nerve impulse reaches the end of the pre-synaptic neuron.
2. Depolarization of the pre-synaptic membrane causes voltage-gated calcium channels to open. Calcium ions diffuse into the pre-synaptic neuron.
3. Influx of calcium causes vesicles of neurotransmitter to move to the pre-synaptic membrane and fuse with it, releasing the neurotransmitter into the synaptic cleft by exocytosis.
4. The neurotransmitter diffuses across the synaptic cleft and binds to receptors in the post-synaptic membrane.
5. The receptors are transmitter-gated ion channels, which open when neurotransmitter binds. Sodium and other positively charged ions diffuse into the post-synaptic neuron. This causes depolarization of the post-synaptic membrane.
6. The depolarization passes on down the post-synaptic neuron as an action potential.
7. Neurotransmitter in the synaptic cleft is rapidly broken down, to prevent continuous synaptic transmission. For example, acetylcholine is broken down by cholinesterase in synapses that use it as a neurotransmitter. Calcium ions are pumped out of the pre-synaptic neuron into the synaptic cleft.

The figure (right) shows the events that occur during synaptic transmission.

Stages in synaptic transmission

③ Vesicles of neurotransmitter move to the membrane and release their contents

① Nerve impulse reaches the end of the pre-synaptic neuron

⑦ Calcium is pumped out. Neurotransmitter is broken down in the cleft and reabsorbed into the vesicles

synaptic knob

vesicles of neuro-transmitter

② Calcium diffuses in through calcium channels

④ Neuro-transmitter diffuses across the synaptic cleft and binds to receptors

⑤ Sodium ions enter the post-synaptic neuron and cause depolarization

⑥ Nerve impulse setting off along the post-synaptic neuron

Nerve impulses

RESTING POTENTIALS

The resting potential is the electrical potential across the plasma membrane of a cell that is not conducting an impulse. Neurons pump ions across their plasma membranes by active transport. Sodium is pumped out of the neuron and potassium is pumped in. Concentration gradients of both sodium and potassium are established across the membrane. The inside of the neuron develops a net negative charge, compared with the outside, because of the presence of chloride and other negatively charged ions. There is therefore an electrical potential or voltage across the membrane. This is called the **resting potential**.

ACTION POTENTIALS

An action potential is the reversal and restoration of the electrical potential across the plasma membrane of a cell, as an electrical impulse passes along it (depolarization and repolarization).

When an impulse passes along the neuron, sodium and potassium ions are allowed to diffuse across the membrane, through voltage-gated ion channels. The electrical potential across the membrane is initially reversed but is then restored. This is called an **action potential**. The figure (right) shows the changes in membrane polarization that occur during an action potential. The way in which action potentials pass down nerve fibres is explained below.

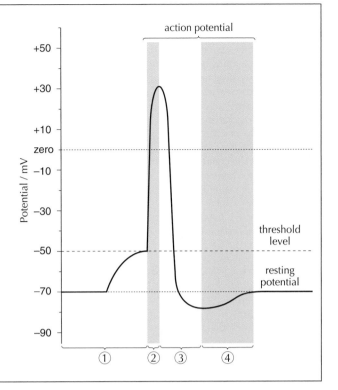

STAGES IN THE PASSAGE OF A NERVE IMPULSE

① An action potential in one part of a neuron causes an action potential to develop in the next section of the neuron. This is due to diffusion of sodium ions between the region with an action potential and the region at the resting potential. These ion movements, local currents, reduce the resting potential. If the potential rises above the threshold level, voltage-gated channels open.

② Sodium channels open very quickly and sodium ions diffuse into the neuron down the concentration gradient. This reduces the membrane potential and causes more sodium channels to open. The entry of positively charged sodium ions causes the inside of the neuron to develop a net positive charge compared to the outside – the potential across the membrane is reversed. This is called **depolarization**.

③ Potassium channels open after a short delay. Potassium ions diffuse out of the neuron down the concentration gradient through the opened channels. The exit of positively charged potassium ions cause the inside of the neuron to develop a net negative charge again compared with the outside – the potential across the membrane is restored. This is called **repolarization**.

④ Concentration gradients of sodium and potassium across the membrane are restored by the active transport of sodium ions out of the neuron and potassium ions into the neuron. This restores the resting potential and the neuron is then ready to conduct another nerve impulse. As before, sodium ions diffuse along inside the neuron from an adjacent region that has already depolarized and initiate depolarization.

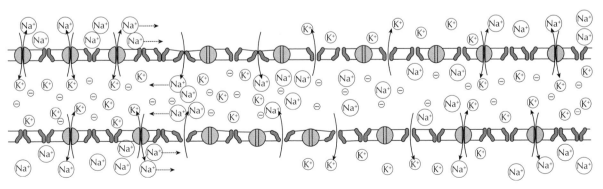

Ion movements during an action potential

Maintaining the internal environment

HOMEOSTASIS

Blood, and tissue fluid derived from blood, flow around or close to all cells in the body. Blood and tissue fluid form the internal environment of the body. This internal environment is controlled and varies very little despite large variations in the external environment. The control process is called **homeostasis**. *Homeostasis is maintaining the internal environment of the body between limits.*

The parameters controlled include
- body temperature
- blood pH
- carbon dioxide concentration
- blood glucose concentration
- water balance

The nervous system and the endocrine system are both involved in controlling the internal environment. The endocrine system consists of glands, which release hormones that are transported in the blood.

The endocrine and nervous systems

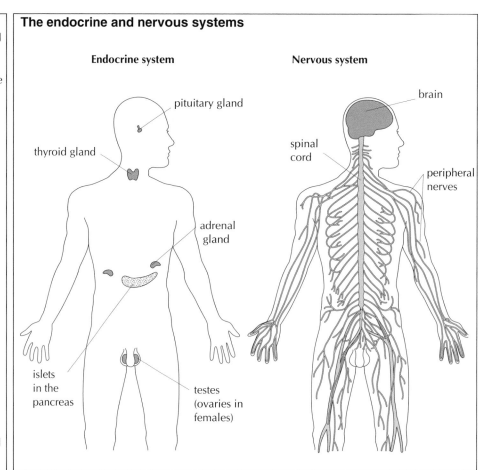

Endocrine system — pituitary gland, thyroid gland, adrenal gland, islets in the pancreas, testes (ovaries in females)

Nervous system — brain, spinal cord, peripheral nerves

CONTROLLING LEVELS BY NEGATIVE FEEDBACK

1. Feedback

In feedback systems, the level of a product feeds back to control the rate of its own production.

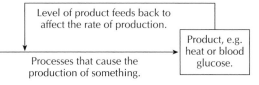

Level of product feeds back to affect the rate of production.

Processes that cause the production of something.

Product, e.g. heat or blood glucose.

2. Negative feedback

Negative feedback has a stabilizing effect because a change in levels always causes the opposite change. A rise in levels feeds back to decrease production and reduce the level. A decrease in levels feeds back to increase production and raise the level. These are both negative feedback.

3. Monitoring levels

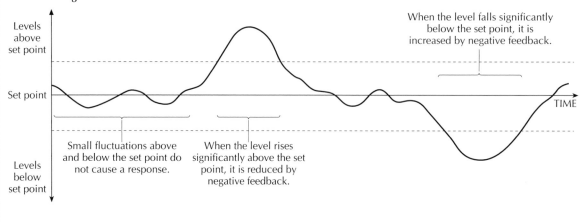

Levels above set point

Set point

Levels below set point

When the level falls significantly below the set point, it is increased by negative feedback.

Small fluctuations above and below the set point do not cause a response.

When the level rises significantly above the set point, it is reduced by negative feedback.

TIME

Body temperature and blood glucose

CONTROL OF BODY TEMPERATURE

The hypothalamus of the brain monitors the temperature of the blood and compares it with a set point, usually close to 37°C. If the blood temperature is lower or higher than the set point the hypothalamus sends messages to parts of the body to make them respond and bring the temperature back to the set point – negative feedback. These messages are carried by neurons. The responses affect the rate at which heat is produced, the rate at which it is transferred between parts of the body in the blood, or the rate at which it is lost from the body.

Responses to overheating	Responses to chilling
Skin arterioles become wider, so more blood flows through the skin. This blood transfers heat from the core of the body to the skin. The temperature of the skin rises, so more heat is lost from it to the environment.	Skin arterioles become narrower and they bring less blood to the skin. The blood capillaries in the skin do not move, but less blood flows through them. The temperature of the skin falls, so less heat is lost from it to the environment.
Skeletal muscles remain relaxed and resting so that they do not generate heat.	Skeletal muscles do many small rapid contractions to generate heat. This is called shivering.
Sweat glands secrete large amounts of sweat making the surface of the skin damp. Water evaporates from the damp skin and this has a cooling effect.	Sweat glands do not secrete sweat and the skin remains dry.

CONTROL OF BLOOD GLUCOSE

Blood glucose concentration cannot be kept as steady as body temperature. Instead it is usually kept between 4 and 8 millimoles per dm^3 of blood. Cells in the pancreas monitor the concentration and send hormone messages to target organs when the level is low or high. Responses by the target organs affect the rate at which glucose is loaded into the blood or unloaded from it. The mechanisms involved are another example of negative feedback.

Responses to high blood glucose levels	Responses to low blood glucose levels
β cells in the pancreatic islets produce insulin.	α cells in the pancreatic islets produce glucagon.
Insulin stimulates the liver and muscle cells to absorb glucose from the blood and convert it to glycogen. Granules of glycogen are stored in the cytoplasm of these cells. Other cells are stimulated to absorb glucose and use it in cell respiration instead of fat. These processes lower the blood glucose level.	Glucagon stimulates liver cells to break glycogen down into glucose and release the glucose into the blood. This raises the blood glucose level.

DIABETES

In some people the control of blood glucose does not work effectively and the concentration can rise or fall beyond the normal limits. The full name for this condition is **diabetes mellitus**. There are two forms of this condition, which are compared in the table below:

Type I diabetes	Type II diabetes
The onset is usually during childhood.	The onset is usually after childhood.
β cells produce insufficient insulin.	Target cells become insensitive to insulin.
Insulin injections are used to control glucose levels.	Insulin injections are not usually needed.
Diet cannot by itself control the condition.	Low carbohydrate diets usually control the condition.

Reproductive systems

THE FEMALE REPRODUCTIVE SYSTEM

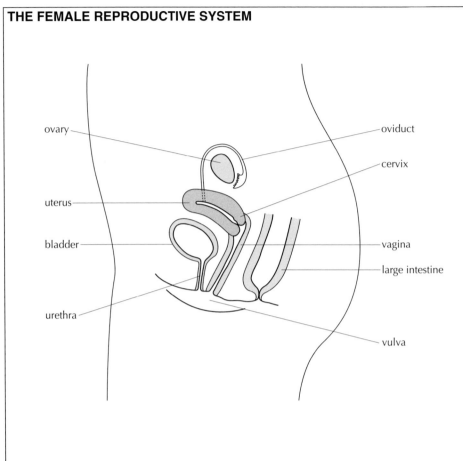

- ovary
- uterus
- bladder
- urethra
- oviduct
- cervix
- vagina
- large intestine
- vulva

FEMALE SEX HORMONES

The pituitary gland produces FSH and LH. These two hormones affect processes in the ovary. FSH stimulates the development of follicles –fluid filled sacs that contain an egg cell.

LH stimulates follicles to become mature, release their egg (ovulation) and then develop into a structure called the corpus luteum.

The ovary produces estrogen and progesterone. These two hormones stimulate the development of female secondary sexual characteristics during puberty. They also stimulate the development of the uterus lining that is needed during pregnancy. Unless a woman is pregnant the levels of the female sex hormones rise and fall according to a cycle, which is described on page 57.

TESTOSTERONE

Cells in the testes of males produce testosterone – the male sex hormone. Testosterone has several roles.

- The developing testes of a male fetus secrete testosterone, which causes male genitalia, including a penis, to develop in the fetus
- Levels of testosterone rise during puberty and cause male secondary sexual characteristics to develop – pubic hair, an enlarged penis and growth of skeletal muscles for example
- During adulthood, testosterone maintains the sex drive, the instinct which encourages men to have sexual intercourse and therefore pass on their genes to offspring.

Testosterone is also one of the hormones needed to stimulate sperm production by the testes.

THE MALE REPRODUCTIVE SYSTEM

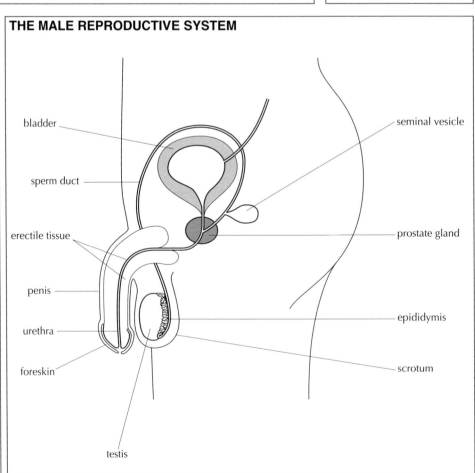

- bladder
- sperm duct
- erectile tissue
- penis
- urethra
- foreskin
- testis
- seminal vesicle
- prostate gland
- epididymis
- scrotum

The menstrual cycle

Between puberty and the menopause, women who are not pregnant follow a cycle called the menstrual cycle. This cycle is controlled by hormones FSH and LH produced by the pituitary gland and estrogen and progesterone produced by the ovary. The figure below shows the levels of these hormones during the menstrual cycle. It also shows the changes in the ovary and in the uterus.

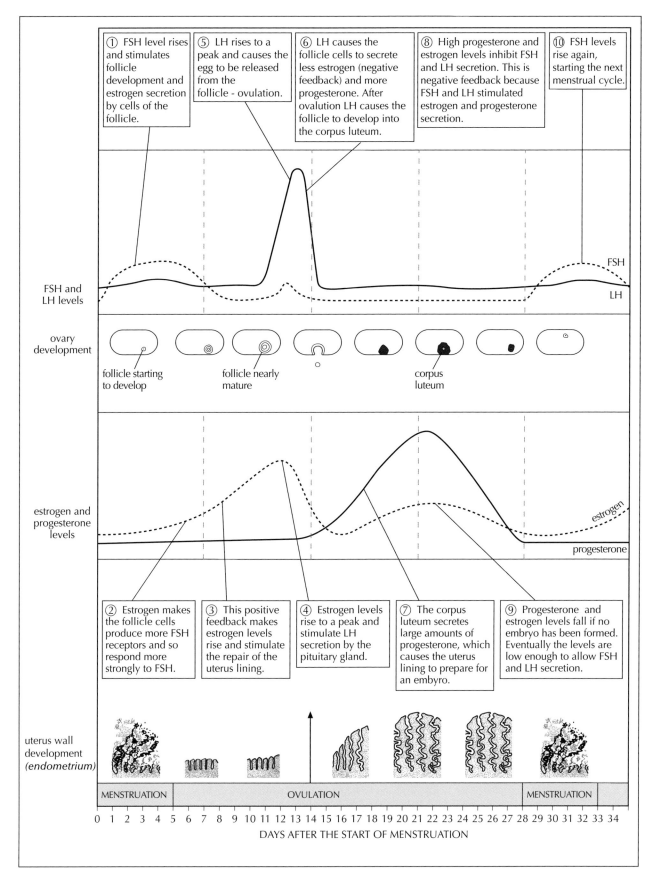

① FSH level rises and stimulates follicle development and estrogen secretion by cells of the follicle.

⑤ LH rises to a peak and causes the egg to be released from the follicle - ovulation.

⑥ LH causes the follicle cells to secrete less estrogen (negative feedback) and more progesterone. After ovulation LH causes the follicle to develop into the corpus luteum.

⑧ High progesterone and estrogen levels inhibit FSH and LH secretion. This is negative feedback because FSH and LH stimulated estrogen and progesterone secretion.

⑩ FSH levels rise again, starting the next menstrual cycle.

FSH and LH levels

FSH

LH

ovary development

follicle starting to develop

follicle nearly mature

corpus luteum

estrogen and progesterone levels

estrogen

progesterone

② Estrogen makes the follicle cells produce more FSH receptors and so respond more strongly to FSH.

③ This positive feedback makes estrogen levels rise and stimulate the repair of the uterus lining.

④ Estrogen levels rise to a peak and stimulate LH secretion by the pituitary gland.

⑦ The corpus luteum secretes large amounts of progesterone, which causes the uterus lining to prepare for an embyro.

⑨ Progesterone and estrogen levels fall if no embryo has been formed. Eventually the levels are low enough to allow FSH and LH secretion.

uterus wall development (endometrium)

MENSTRUATION | OVULATION | MENSTRUATION

0 1 2 3 4 5 6 7 8 9 10 11 12 13 14 15 16 17 18 19 20 21 22 23 24 25 26 27 28 29 30 31 32 33 34
DAYS AFTER THE START OF MENSTRUATION

In vitro fertilization

INFERTILITY

Some couples do not achieve fertilization and pregnancy when they wish to, despite sexual intercourse during the period in the middle of the menstrual cycle when ovulation usually occurs. This is called infertility. It may be temporary, because the causes can be resolved, or permanent.

Approximately one in six couples have some experience of temporary or permanent infertility. Many of these couples can be helped to have a child by *in vitro* fertilization – IVF. For example, blocked oviducts in a woman prevent conception, but IVF can overcome this problem. Other problems cannot be resolved by IVF, for example low or zero sperm counts in men.

The process of IVF is outlined (right).

Timetable for IVF

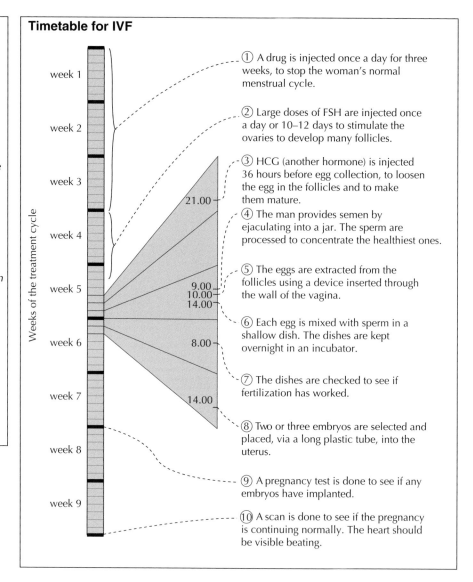

Weeks of the treatment cycle

week 1
week 2
week 3
week 4
week 5
week 6
week 7
week 8
week 9

21.00
9.00
10.00
14.00
8.00
14.00

① A drug is injected once a day for three weeks, to stop the woman's normal menstrual cycle.

② Large doses of FSH are injected once a day or 10–12 days to stimulate the ovaries to develop many follicles.

③ HCG (another hormone) is injected 36 hours before egg collection, to loosen the egg in the follicles and to make them mature.

④ The man provides semen by ejaculating into a jar. The sperm are processed to concentrate the healthiest ones.

⑤ The eggs are extracted from the follicles using a device inserted through the wall of the vagina.

⑥ Each egg is mixed with sperm in a shallow dish. The dishes are kept overnight in an incubator.

⑦ The dishes are checked to see if fertilization has worked.

⑧ Two or three embryos are selected and placed, via a long plastic tube, into the uterus.

⑨ A pregnancy test is done to see if any embryos have implanted.

⑩ A scan is done to see if the pregnancy is continuing normally. The heart should be visible beating.

ETHICAL ISSUES ASSOCIATED WITH IVF

Some issues are controversial and around the world the views held by people may vary considerably. Ethical issues involve questioning whether something is wrong or right. Decisions cannot be made using scientific methods, but scientists have an obligation to consider ethical issues.

Ethical arguments against IVF

- Inherited forms of infertility might be passed on to children, which means that the suffering of the parents is repeated in their offspring.
- More embryos are often produced than are needed and the spare embryos are sometimes killed, denying them the chance of life.
- Embryologists select embryos to transfer to the uterus, so humans are deciding whether new individuals survive or die.
- IVF is an unnatural process, carried out in laboratories, in contrast to natural conception occurring as a result of an act of love.
- Infertility should be accepted as the will of God and it is wrong to try to circumvent it by using IVF to have a child.

Ethical arguments for IVF

- Many forms of infertility are due to environmental factors, so offspring will not inherit them.
- Any embryos that are killed during IVF are unable to feel pain or suffer, because their nervous system has not developed.
- Suffering due to genetic disease could be reduced if embryos were screened before being transferred to the uterus.
- Parents willing to go through the process of IVF must have a strong desire for children and so are likely to be loving parents.
- Infertility brings great unhappiness to parents who want to have children, which in some cases can be overcome by IVF.

EXAM QUESTIONS ON TOPIC 6

1 Respiration in humans and other mammals generates heat which can be used to keep the body temperature above that of the surroundings.

Many mammals found in the southern hemisphere, including marsupials, vary their body temperature according to a daily cycle. The mouse lemur (*Microcebus myoxinus*) is an example of such a mammal. To investigate this daily cycle, *M. myoxinus* was studied in its native habitat in Madagascar. Data-loggers which recorded body temperature (T_b) over 24-hour periods were implanted in the bodies of several of these mammals. Air temperature (T_a) was recorded at the same time. A typical set of results is shown in the graph below.

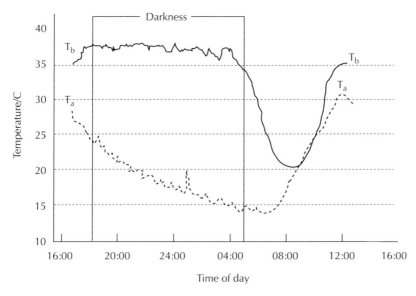

[Source: Cossins and Barnes, Nature (1996), 384, page 582]

a) Using only the data in the graph, state two differences between T_a and T_b during the hours of darkness.   [2]

b) T_b rises from 08:00 to 12:00. Explain briefly how this temperature rise occurs.   [2]

c) Predict, with a reason, whether *M. myoxinus* is active in the hours of daylight or the hours of darkness.   [1]

2 a) (i) State the function of phagocytic leukocytes.   [1]

   (ii) Outline where in the body phagocytic leukocytes carry out their function.   [2]

   b) Explain briefly the need for small numbers of many types of B-lymphocyte in the body.   [2]

3 The diagram right shows part of the human gas exchange system.

   a) State the name of the parts labelled I and II.   [2]

   b) I and II allow the lungs to be ventilated. Explain briefly the need for ventilation.   [2]

   c) Draw and label a diagram of alveoli.   [3]

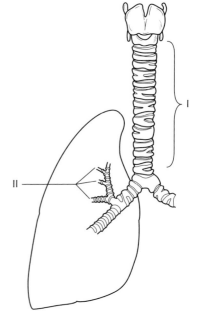

DNA structure and replication

DNA STRUCTURE

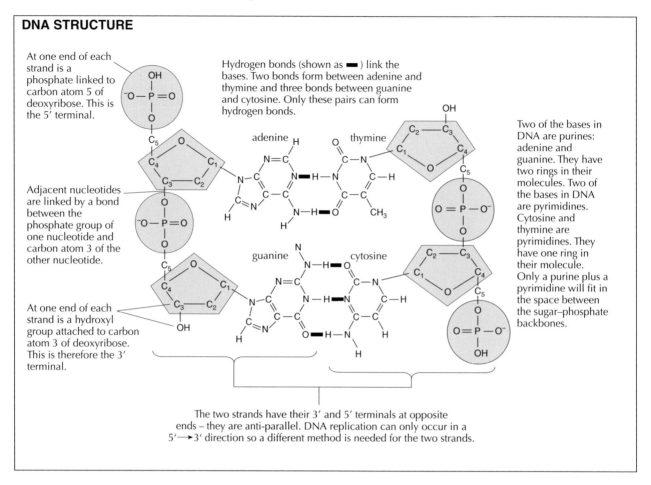

At one end of each strand is a phosphate linked to carbon atom 5 of deoxyribose. This is the 5′ terminal.

Hydrogen bonds (shown as ■) link the bases. Two bonds form between adenine and thymine and three bonds between guanine and cytosine. Only these pairs can form hydrogen bonds.

Two of the bases in DNA are purines: adenine and guanine. They have two rings in their molecules. Two of the bases in DNA are pyrimidines. Cytosine and thymine are pyrimidines. They have one ring in their molecule. Only a purine plus a pyrimidine will fit in the space between the sugar–phosphate backbones.

Adjacent nucleotides are linked by a bond between the phosphate group of one nucleotide and carbon atom 3 of the other nucleotide.

At one end of each strand is a hydroxyl group attached to carbon atom 3 of deoxyribose. This is therefore the 3′ terminal.

The two strands have their 3′ and 5′ terminals at opposite ends – they are anti-parallel. DNA replication can only occur in a 5′⟶3′ direction so a different method is needed for the two strands.

DNA REPLICATION

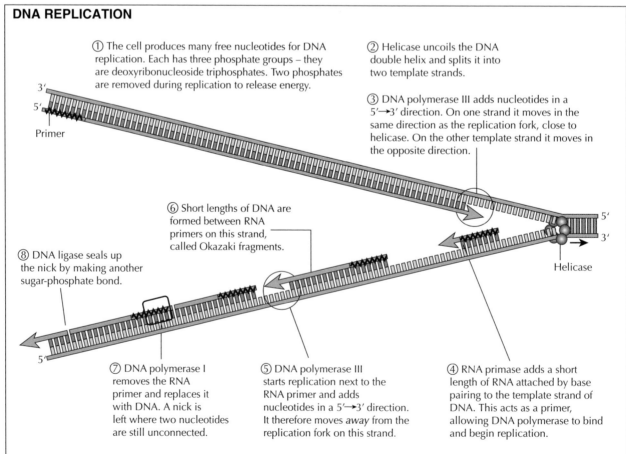

① The cell produces many free nucleotides for DNA replication. Each has three phosphate groups – they are deoxyribonucleoside triphosphates. Two phosphates are removed during replication to release energy.

② Helicase uncoils the DNA double helix and splits it into two template strands.

③ DNA polymerase III adds nucleotides in a 5′⟶3′ direction. On one strand it moves in the same direction as the replication fork, close to helicase. On the other template strand it moves in the opposite direction.

⑥ Short lengths of DNA are formed between RNA primers on this strand, called Okazaki fragments.

⑧ DNA ligase seals up the nick by making another sugar-phosphate bond.

⑦ DNA polymerase I removes the RNA primer and replaces it with DNA. A nick is left where two nucleotides are still unconnected.

⑤ DNA polymerase III starts replication next to the RNA primer and adds nucleotides in a 5′⟶3′ direction. It therefore moves *away* from the replication fork on this strand.

④ RNA primase adds a short length of RNA attached by base pairing to the template strand of DNA. This acts as a primer, allowing DNA polymerase to bind and begin replication.

Organization of DNA in eukaryotes

DNA IN PROKARYOTES AND EUKARYOTES

All eukaryotes and prokaryotes use DNA as their genetic material and use the same genetic code, but there are differences in the way that the DNA is organized and used. Prokaryotes have naked DNA, which consists mostly of single copy genes that are transcribed and translated without modification. The situation in eukaryotes is more complex:

Replication initiation sites

Replication of DNA begins at special initiation points. Eukaryotes have many of these initiation points along each chromosome. Most prokaryotes have only one point on their DNA molecule where replication is initiated.

Nucleosomes

In eukaryotes, the DNA is associated with proteins to form nucleosomes – globular structures that contain eight histone proteins, with DNA wrapped around. Another histone protein bonds the structure together (above right). In an interphase nucleus in eukaryotes the DNA resembles a string of beads (right). Nucleosomes have two functions:
- they help to package up the DNA during mitosis and meiosis by the process of supercoiling
- they can be used to mark particular genes, either to promote gene expression by transcription and translation, or to cause silencing of a gene by preventing transcription.

Repetitive sequences

Much of the DNA in eukaryotes consists of repetitive base sequences, which are not translated. Highly repetitive sequences, sometimes called satellite DNA, are sequences of between 5 and 300 bases, that may be repeated as many as 10 000 times. These constitute 5–45% of typical eukaryote DNA. Its function is not yet clear.
A surprisingly small proportion of eukaryotic DNA is single copy, or unique genes.

Introns and exons

Many genes in eukaryotes contain **introns** – sequences of bases that are transcribed, but not translated. **Exons** are sequences of bases that are transcribed and translated. A typical eukaryote gene consists of a series of exons and introns. After transcription of the whole gene, the introns are removed to form mature mRNA, in a process called post-transcriptional modification (below right). Prokaryotes do not usually have introns in their genes.

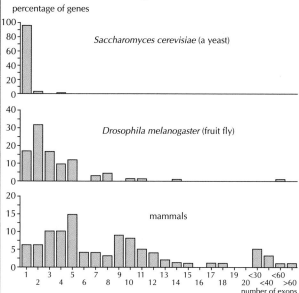

Nucleosome structure

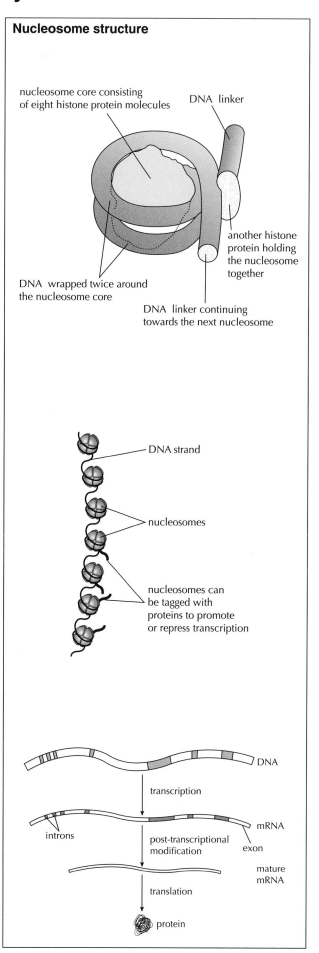

Transcription of DNA

RNA POLYMERASE AND TRANSCRIPTION

DNA is split into two strands by RNA polymerase. One of these strands forms the template for transcription. The base sequence of the mRNA is complementary to it. The other strand has the same base sequence as the mRNA (except for T instead of U) and is therefore called the **sense strand**. The strand that forms the template and is transcribed is called the **antisense strand**.

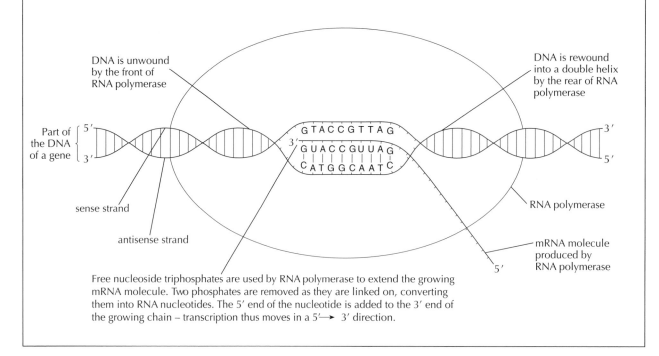

DNA is unwound by the front of RNA polymerase

DNA is rewound into a double helix by the rear of RNA polymerase

Part of the DNA of a gene

sense strand

antisense strand

RNA polymerase

mRNA molecule produced by RNA polymerase

Free nucleoside triphosphates are used by RNA polymerase to extend the growing mRNA molecule. Two phosphates are removed as they are linked on, converting them into RNA nucleotides. The 5′ end of the nucleotide is added to the 3′ end of the growing chain – transcription thus moves in a 5′⟶ 3′ direction.

TRANSLATING THE GENETIC CODE

Messenger RNA carries the information needed for making polypeptides out from the nucleus to the cytoplasm of eukaryotic cells. The information is in a coded form, which is decoded during translation. The base sequence of mRNA is translated into the amino acid sequence of a polypeptide. Key features of the code are described (below left). The meaning of each codon is shown in the table (below right).

The genetic code is a **triplet code** – three bases code for one amino acid. A group of three bases is called a **codon**. There are 64 different codons (4^3). This gives more than enough codons to code for the twenty amino acids in proteins. If codons consisted of two bases there would be sixteen (4^2) – not enough. None of the 64 codons are unused. Instead, the genetic code is **degenerate**. This means that it is possible for two or more codons to code for the same amino acid. The genetic code is **universal**. With just a few minor exceptions, living organisms use precisely the same code. Viruses also use this code.

First base of codon (5′ end)	Second base of codon on messenger RNA				Third base of codon (3′ end)
	U	C	A	G	
U	Phenylalanine	Serine	Tyrosine	Cysteine	U
	Phenylalanine	Serine	Tyrosine	Cysteine	C
	Leucine	Serine	STOP	STOP	A
	Leucine	Serine	STOP	Tryptophan	G
C	Leucine	Proline	Histidine	Arginine	U
	Leucine	Proline	Histidine	Arginine	C
	Leucine	Proline	Glutamine	Arginine	A
	Leucine	Proline	Glutamine	Arginine	G
A	Isoleucine	Threonine	Asparagine	Serine	U
	Isoleucine	Threonine	Asparagine	Serine	C
	Isoleucine	Threonine	Lysine	Arginine	A
	Methionine / START	Threonine	Lysine	Arginine	G
G	Valine	Alanine	Aspartic acid	Glycine	U
	Valine	Alanine	Aspartic acid	Glycine	C
	Valine	Alanine	Glutamic acid	Glycine	A
	Valine	Alanine	Glutamic acid	Glycine	G

Ribosomes and transfer RNA

tRNA AND tRNA ACTIVATING ENZYMES

There are many different types of tRNA in a cell, which have an important role in translation.

All tRNA molecules have:
- sections that become double stranded by base pairing, creating loops (above right)
- a triplet of bases called the anticodon, in a loop of seven bases
- two other loops
- the base sequence CCA at the 3′ terminal, which forms a site for attaching an amino acid.

These features allow all tRNA molecules to bind to three sites on the ribosome.

The base sequence of tRNA molecules varies and this causes some variable features in its structure:
- an extra small loop is sometimes present
- the base paired sections are sometimes helical.

The variable features give each type of tRNA a distinctive three-dimensional shape and distinctive chemical properties (below right). This allows the correct amino acid to be attached to the 3′ terminal by an enzyme called a **tRNA activating enzyme**. There are twenty different tRNA activating enzymes – one for each of the twenty different amino acids. Each of these enzymes attaches one particular amino acid to all of the tRNA molecules that have an anticodon corresponding to that amino acid. The tRNA activating enzymes recognize these tRNA molecules by their shape and chemical properties.

Energy from ATP is needed for the attachment of amino acids. A high-energy bond is created between the amino acid and the tRNA. Energy from this bond is later used to link the amino acid to the growing polypeptide chain during translation.

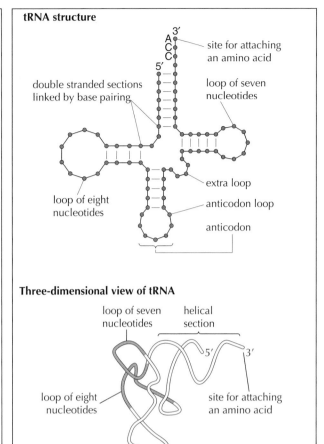

tRNA structure

Three-dimensional view of tRNA

STRUCTURE AND FUNCTION OF RIBOSOMES

Ribosomes have a complex structure, with these features.
- Proteins and ribosomal RNA molecules (rRNA) both form part of the structure.
- There are two subunits, one large and one small.
- There are three binding sites for tRNA on the surface of the ribosome. Two tRNA molecules can bind at the same time to the ribosome.
- There is a binding site for mRNA on the surface of the ribosome.

The structure of a ribosome is shown in outline in the figure (right), with the three tRNA binding sites.

Ribosomes in the cytoplasm are called **free ribosomes**. They synthesize proteins for use within the cell. Ribosomes can also be attached to membranes of the endoplasmic reticulum. They are called **bound ribosomes** and synthesize proteins for secretion from the cell or for lysosomes.

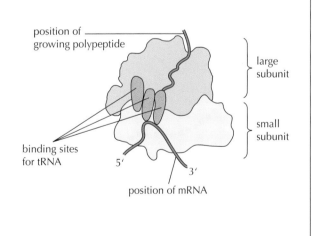

PEPTIDE BONDS

Ribosomes are the site of polypeptide synthesis. This involves linking amino acids together by a condensation reaction (shown on page 15). The linkage between the amino acids is a peptide bond. Perhaps unexpectedly, it is rRNA and not proteins in the ribosome that catalyse the reaction in which the peptide bond is formed. The diagram (right) shows a peptide bond between two amino acids.

peptide bond

Polysomes and polypeptide elongation

The figure (right) is an electron micrograph showing groups of ribosomes called **polysomes** (or polyribosomes). A polysome is a group of ribosomes moving along the same mRNA, as they simultaneously translate it. Each ribosome follows a series of steps that is repeated many times to translate the mRNA. One amino acid is added to the elongating polypeptide each time the cycle of steps is repeated (see below). As ribosomes move along the mRNA towards the 3' end, the polypeptide is gradually elongated.

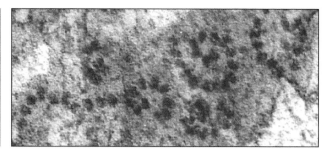

(× 180 000)

POLYPEPTIDE ELONGATION

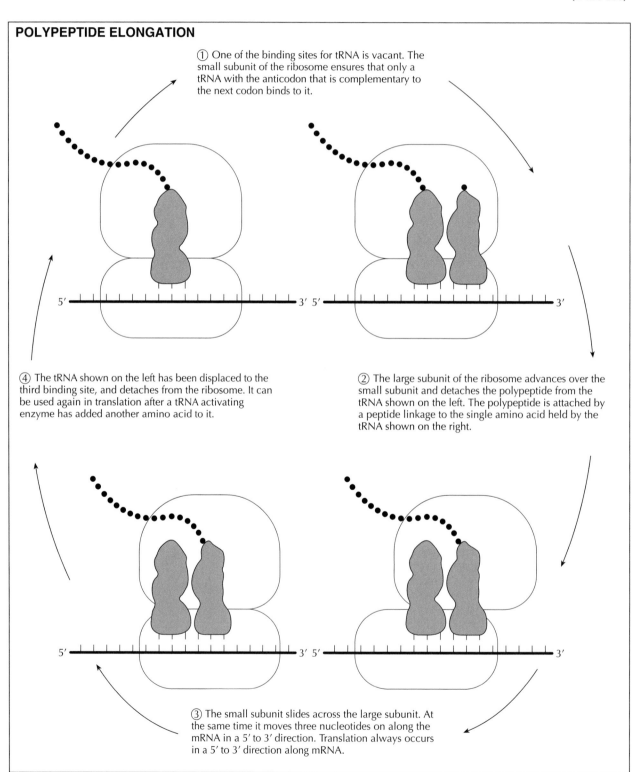

① One of the binding sites for tRNA is vacant. The small subunit of the ribosome ensures that only a tRNA with the anticodon that is complementary to the next codon binds to it.

④ The tRNA shown on the left has been displaced to the third binding site, and detaches from the ribosome. It can be used again in translation after a tRNA activating enzyme has added another amino acid to it.

② The large subunit of the ribosome advances over the small subunit and detaches the polypeptide from the tRNA shown on the left. The polypeptide is attached by a peptide linkage to the single amino acid held by the tRNA shown on the right.

③ The small subunit slides across the large subunit. At the same time it moves three nucleotides on along the mRNA in a 5' to 3' direction. Translation always occurs in a 5' to 3' direction along mRNA.

Starting and stopping translation

Special steps are needed to start the process of translation and to stop it. These steps are called **initiation** and **termination**. The three stages of translation are thus initiation, elongation and termination.

INITIATION OF TRANSLATION

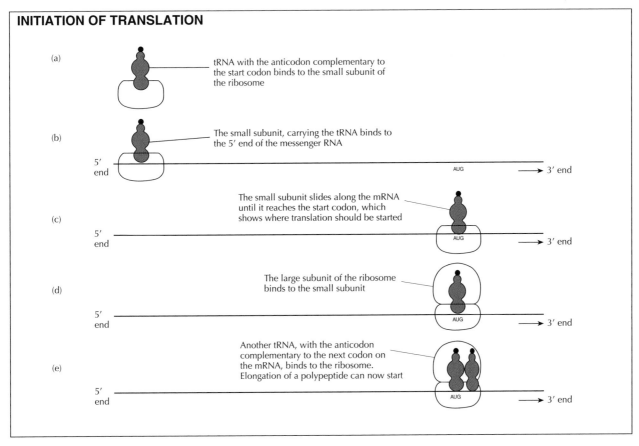

(a) tRNA with the anticodon complementary to the start codon binds to the small subunit of the ribosome

(b) The small subunit, carrying the tRNA binds to the 5′ end of the messenger RNA

(c) The small subunit slides along the mRNA until it reaches the start codon, which shows where translation should be started

(d) The large subunit of the ribosome binds to the small subunit

(e) Another tRNA, with the anticodon complementary to the next codon on the mRNA, binds to the ribosome. Elongation of a polypeptide can now start

TERMINATION OF TRANSLATION

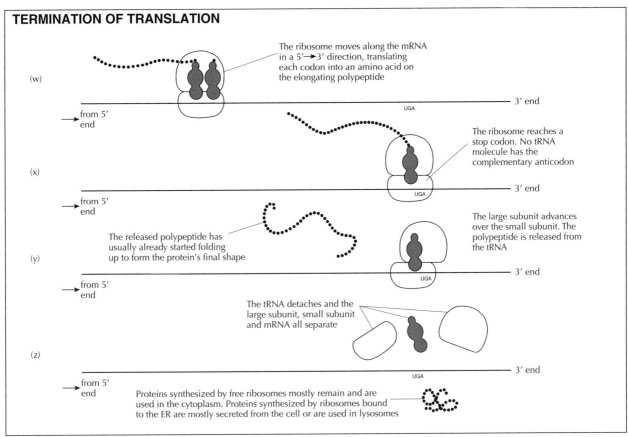

(w) The ribosome moves along the mRNA in a 5′→3′ direction, translating each codon into an amino acid on the elongating polypeptide

(x) The ribosome reaches a stop codon. No tRNA molecule has the complementary anticodon

(y) The large subunit advances over the small subunit. The polypeptide is released from the tRNA

The released polypeptide has usually already started folding up to form the protein's final shape

(z) The tRNA detaches and the large subunit, small subunit and mRNA all separate

Proteins synthesized by free ribosomes mostly remain and are used in the cytoplasm. Proteins synthesized by ribosomes bound to the ER are mostly secreted from the cell or are used in lysosomes

Nucleic acids and proteins 65

Intramolecular bonding in proteins

Polypeptides have a main chain consisting of a repeating sequence of covalently bonded carbon and nitrogen atoms: N – C – C – N – C – C, and so on. Each nitrogen atom has a hydrogen atom bonded to it (N – H). Every second carbon atom has an oxygen atom bonded to it (C = O).

$$
\begin{array}{c}
\quad\quad\; H \quad\;\; O \\
\quad\quad\; | \quad\quad\; || \\
N - C - C - N - C - C \\
\quad | \quad\quad\; || \\
\quad H \quad\quad\; O
\end{array}
$$

Hydrogen bonds can form between N – H and C = O groups, if they are brought close together. For example, if sections of polypeptide run parallel, hydrogen bonds can form between them. The structure that develops is called a ß-pleated sheet. If the polypeptide is wound into a right-handed helix, hydrogen bonds can form between adjacent turns of the helix. The structure that develops is called an α-helix. Because the groups forming hydrogen bonds are regularly spaced, secondary structures always have the same dimensions.

In addition to the hydrogen bonding in ß-pleated sheets and α-helices, there are many other types of bonding. Most of these involve the R groups of the amino acids. The figure (below) shows some of these bonds.

β-PLEATED SHEET

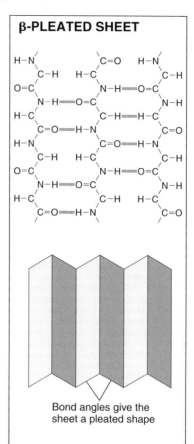

Bond angles give the sheet a pleated shape

α -HELIX

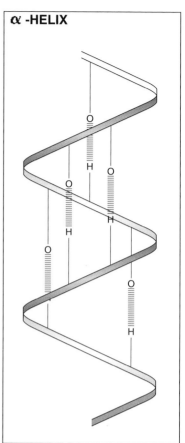

Types of intramolecular bond in proteins

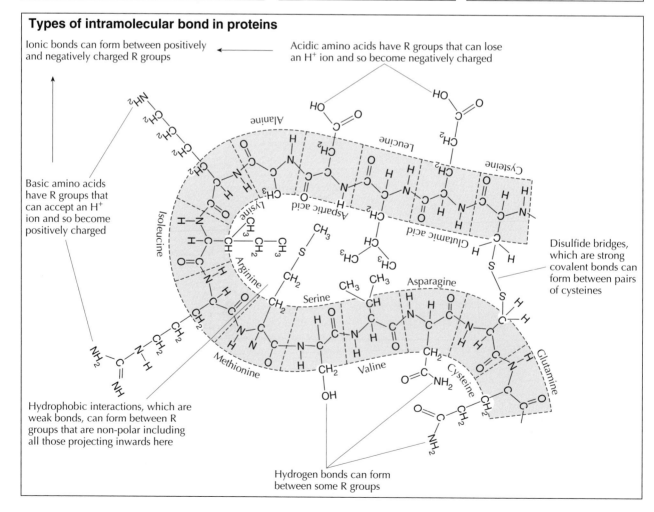

Ionic bonds can form between positively and negatively charged R groups

Acidic amino acids have R groups that can lose an H⁺ ion and so become negatively charged

Basic amino acids have R groups that can accept an H⁺ ion and so become positively charged

Disulfide bridges, which are strong covalent bonds can form between pairs of cysteines

Hydrophobic interactions, which are weak bonds, can form between R groups that are non-polar including all those projecting inwards here

Hydrogen bonds can form between some R groups

Protein structure

Proteins have a complex structure, which can be explained by defining four levels of structure, primary, secondary, tertiary and quaternary structure.

PRIMARY STRUCTURE

Primary structure is the number and sequence of amino acids in a polypeptide. Most polypeptides consist of between 50 and 1000 amino acids. The primary structure is determined by the base sequence of the gene that codes for the polypeptide. The figure (below) shows the primary structure of ß-endorphin, a protein consisting of a single polypeptide of 31 amino acids that acts as a neurotransmitter in the brain.

Primary structure of ß-endorphin

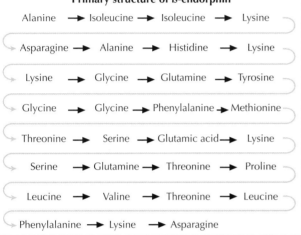

Alanine → Isoleucine → Isoleucine → Lysine

Asparagine → Alanine → Histidine → Lysine

Lysine → Glycine → Glutamine → Tyrosine

Glycine → Glycine → Phenylalanine → Methionine

Threonine → Serine → Glutamic acid → Lysine

Serine → Glutamine → Threonine → Proline

Leucine → Valine → Threonine → Leucine

Phenylalanine → Lysine → Asparagine

SECONDARY STRUCTURE

Secondary structures are regular repeating structures, including ß-pleated sheets and α-helices stabilized by hydrogen bonds between groups in the main chain of the polypeptide. In many proteins, parts of the polypeptide form secondary structures and other parts do not. In some proteins secondary structures do not form at all. In a few proteins almost all of the polypeptide forms secondary structures. For example almost all of myosin molecules is α-helix and almost all of fibroin (silk protein) is β-pleated sheet.
The figure (below) shows the position of secondary structures in lysozyme, using the ribbon model. Sections of α-helix are represented by helical ribbons and sections of β-pleated sheet are represented by arrows.

Ribbon model of lysozyme

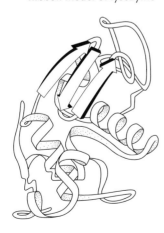

TERTIARY STRUCTURE

Tertiary structure is the three-dimensional conformation of a polypeptide. It is formed when the polypeptide folds up after being produced by translation. The conformation is stabilized by intramolecular bonds that form between amino acids in the polypeptide, especially between their R groups. These include ionic bonds, hydrogen bonds, hydrophobic interactions and disulfide bridges. The intramolecular bonds are often formed between amino acids that are widely separated in the primary structure of the polypeptide, but which are brought together during the folding process. The figure below shows the tertiary structure of lysozyme using the sausage model.

Sausage model of lysozyme

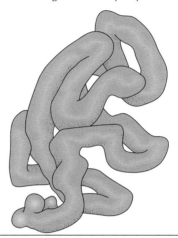

QUATERNARY STRUCTURE

Quaternary structure is the linking together of two or more polypeptides to form a single protein. For example, insulin consists of two polypeptides linked together, collagen consists of three polypeptides and hemoglobin consists of four. In some cases proteins also contain a non-polypeptide structure called a **prosthetic group**. Each of the four polypeptides in hemoglobin is linked to a heme group, which is not made of amino acids. Proteins with a prosthetic group are called **conjugated proteins**.
The figure (below) shows the quaternary structure of hemoglobin.

Sausage model of hemoglobin

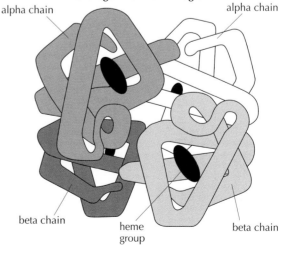

alpha chain   alpha chain

beta chain   heme   beta chain
               group

Protein functions

FUNCTIONS OF FIBROUS AND GLOBULAR PROTEINS

Proteins can be divided into two types according to their shape – **fibrous** or **globular**. Fibrous proteins have a long and narrow shape. They are mostly insoluble in water. Globular proteins have a rounded shape. They are mostly soluble in water. Two examples of both fibrous and globular proteins are given in the table (below).

Proteins have a huge range of functions in living organisms. Some proteins are located in membranes – their functions are listed on page 8. Four of the functions of non-membrane proteins are listed in the table (below). Proteins can also be used as food stores, for example casein in milk, as pigments, for example opsin in the retina, as toxins as in some snake venom, as hormones, for example insulin, and as enzymes.

Function	Example	Details	Shape
Structural	Collagen	The function of collagen is to strengthen bone, tendon and skin. These tissues all produce tough collagen fibres in the spaces between their cells	Fibrous
Transport	Hemoglobin	The function of hemoglobin is to bind oxygen in the lungs and to transport it to respiring tissues	Globular
Movement	Myosin	The function of myosin (with another protein called actin) is to cause contraction in muscle fibres and as a result cause movement in animals	Fibrous
Defence	Immunoglobulin	The function of immunoglobulin is to act as antibodies. Part of the immunoglobulin molecule can be varied, so that an almost endless variety of different antibodies can be produced	Globular

POLAR AND NON-POLAR AMINO ACIDS IN PROTEINS

Amino acids can be divided into two types according to the chemical characteristics of their R group. Polar amino acids have hydrophilic R groups and non-polar amino acids have hydrophobic R groups. The distribution of polar and non-polar amino acids in a protein molecule influence where the protein is located in a cell and what function it can carry out. The figures (below) show examples of this.

Positions of proteins in and out of membranes

Non-polar amino acids in the centre of water-soluble proteins stabilize their structure.

Polar amino acids on the surface of proteins make them water soluble.

Non-polar amino acids cause proteins to remain embedded in membranes.

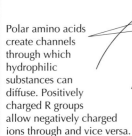

Polar amino acids create channels through which hydrophilic substances can diffuse. Positively charged R groups allow negatively charged ions through and vice versa.

Polar amino acids cause parts of membrane proteins to protrude from the membrane. Transmembrane proteins have two such regions.

Superoxide dismutase – an enzyme found in all aerobic organisms

A ring of amino acids with negatively charged R-groups repel the negatively charged superoxide ions and help to direct them to the active site.

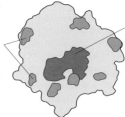

The active site is a cleft containing amino acids with positively charged R-groups which attract the negatively charged superoxide ions that are the substrate of the enzyme.

Lipase – an enzyme that works in the small intestine

polar region

Part of the enzyme molecule acts as a hinged lid which can cover the active site when not in use, hiding the non-polar R-groups.

non-polar region

The active site is a cleft containing amino acids with non-polar R-groups which bind non-polar triglycerides.

A protein cofactor binds to the enzyme, and helps lipase to bind to the surface of lipid droplets because it has non-polar R-groups on its surface.

Enzymes and activation energy

ENERGY CHANGES DURING CHEMICAL REACTIONS

During chemical reactions, reactants are converted into products. Before a molecule of the reactant can take part in the reaction, it has to gain some energy. This is called the **activation energy** of the reaction. The energy is needed to break bonds within the reactant. Later during the progress of the reaction, energy is given out as new bonds are made. Most biological reactions are exothermic – the energy released is greater than the activation energy.

Enzymes reduce the activation energy of the reactions that they catalyse and therefore make it easier for these reactions to occur. The graph (below) shows energy changes during uncatalysed and catalysed exothermic reactions.

The chemical environment provided by the active site for the substrate causes changes within the substrate molecule, which weakens its bonds. The substrate is changed into a transition state, which is different from the transition state during the reaction when an enzyme is not involved. The transition state achieved during binding to the active site has less energy and this is how enzymes are able to reduce the activation energy of reactions.

Energy changes during a chemical reaction

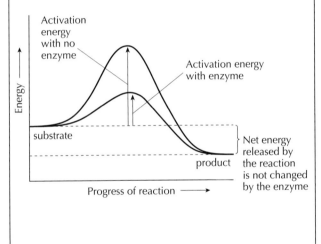

THE INDUCED FIT MODEL

Biochemists have investigated many enzymes and found that the lock and key model does not fully explain the binding of the substrate to the active site. Until the substrate binds, the active site does not fit the substrate precisely. As the substrate approaches the active site and binds to it, the shape of the active site changes and only then does it fit the substrate. The substrate induces the active site to change, weakening bonds in the substrate during the process and thus reducing the activation energy. The figure (right) shows the induced fit model of enzyme activity.

Some enzymes can have quite broad specificity, for example some proteases. The induced fit model explains this better than the lock and key model – if the shape of an active site alters when substrates bind, several different but similar substrates could easily bind successfully to it.

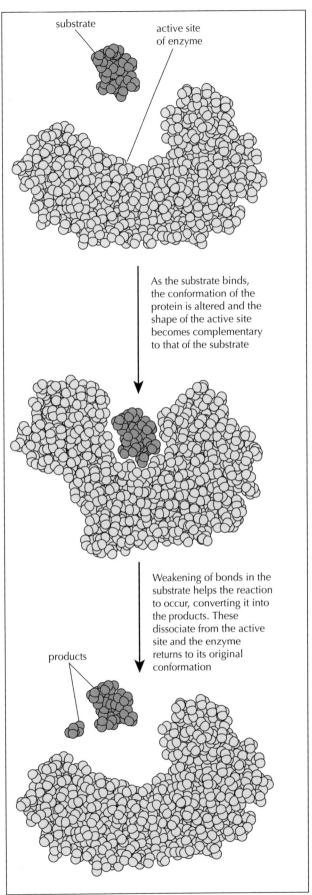

substrate   active site of enzyme

As the substrate binds, the conformation of the protein is altered and the shape of the active site becomes complementary to that of the substrate

Weakening of bonds in the substrate helps the reaction to occur, converting it into the products. These dissociate from the active site and the enzyme returns to its original conformation

products

Enzyme inhibition

Some chemical substances reduce the activity of enzymes or even prevent it completely. These substances are called enzyme inhibitors. Some enzyme inhibitors are **competitive** and some are **non-competitive**. Figures below are a comparison of these types of inhibitor, with an example of each.

Competitive inhibition	Non-competitive inhibition
The substrate and inhibitor are chemically very similar	The substrate and active site are not similar
The inhibitor binds to the active site of the enzyme	The inhibitor binds to the enzyme at a different site from the active site
While the inhibitor occupies the active site, it prevents the substrate from binding and so the activity of the enzyme is prevented until the inhibitor dissociates	The inhibitor changes the conformation of the enzyme. The substrate may still be able to bind, but the active site does not catalyse the reaction, or catalyses it at a slower rate

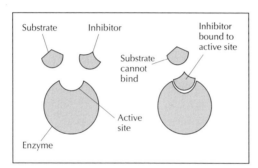

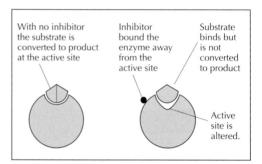

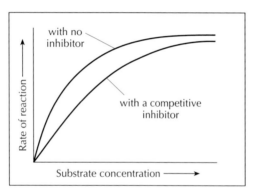

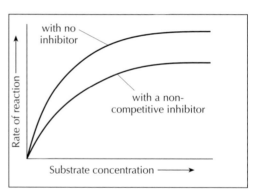

With a fixed low concentration of inhibitor, increases in the substrate concentration gradually reduce the effect of the inhibitor.
The inhibitor and substrate compete for the active site. When the substrate binds to the active site, the inhibitor cannot bind, so the proportion of enzyme molecules that are inhibited becomes less and less. When there are many more substrate molecules than inhibitor molecules, the substrate always wins the competition and binds to the active site. The same maximum enzyme activity rate is then reached as when there is no inhibitor.

With a fixed low concentration of inhibitor, increases in substrate concentration increase enzyme activity. However, the substrate and inhibitor are not competing for the same site. The substrate cannot prevent the binding of the inhibitor, even at very high substrate concentrations. Some of the enzyme molecules therefore remain inhibited and the maximum enzyme activity rate reached is lower than when there is no inhibitor

EXAMPLE

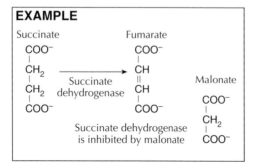

Succinate dehydrogenase is inhibited by malonate

EXAMPLE

Nitric oxide synthase catalyses this reaction:

arginine $\longrightarrow$ citrulline + nitric oxide

Opioids are chemicals that resemble morphine. They are inhibitors of nitric oxide synthase. They do not resemble arginine and bind to a different site on the enzyme, so they are non-competitive inhibitors. Nitric oxide has many signalling roles in human physiology.

Controlling metabolic pathways

METABOLIC PATHWAYS

Metabolic pathways have these features:
- They consist of many chemical reactions that are carried out in a particular sequence.
- An enzyme catalyses each reaction.
- All the reactions occur inside cells.
- Some pathways build up organic compounds (anabolic pathways) and some break them down (catabolic pathways).
- Some metabolic pathways consist of chains of reactions. Glycolysis is an example of a chain of reactions – a chain of ten enzyme-controlled reactions converts glucose into pyruvate.
- Some metabolic pathways consist of cycles of reactions, where a substrate of the cycle is continually regenerated by the cycle. The Krebs cycle is an example.

The figure (opposite) shows the general pattern of reactions in a chain and a cycle.

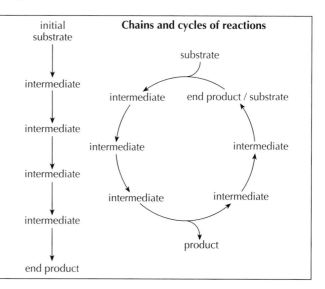

Chains and cycles of reactions

ALLOSTERY AND THE CONTROL OF METABOLIC PATHWAYS

In many metabolic pathways, the product of the last reaction in the pathway inhibits the enzyme that catalyses the first reaction. This is called **end-product inhibition**. The enzyme that is inhibited by the end products is an example of an **allosteric** enzyme. Allosteric enzymes have two non-overlapping binding sites. One of these is the active site. The other is the allosteric site.

In this case the allosteric site is a binding site for the end product. When it binds, the structure of the enzyme is altered so that the substrate is less likely to bind to the active site. This is how the end-product acts as an inhibitor. Binding of the inhibitor is reversible and if it detaches, the enzyme returns to its original conformation, so the active site can bind the substrate easily again (right).

The advantage of this method of controlling metabolic pathways is that if there is an excess of the end-product the whole pathway is switched off and intermediates do not build up. Conversely, as the level of the end-product falls, more and more of the enzymes that catalyse the first reaction will start to work and the whole pathway will become activated. End product inhibition is an example of negative feedback (see example below).

End-product inhibition

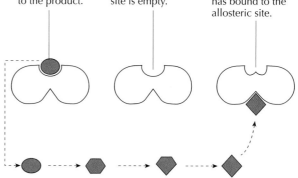

The substrate of the first enzyme in the metabolic pathway is converted by the pathway into an inhibitor of the enzyme.

An example of end product inhibition

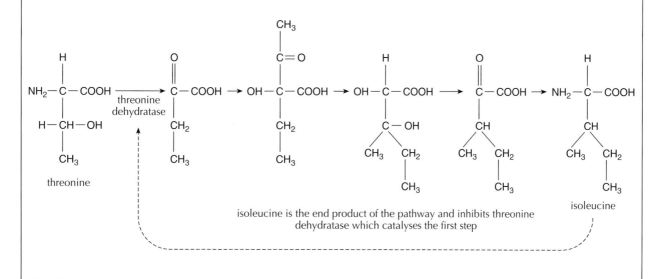

isoleucine is the end product of the pathway and inhibits threonine dehydratase which catalyses the first step

EXAM QUESTIONS ON TOPIC 7

1 An enzyme experiment was conducted at three different temperatures. The graph shows the amount of substrate remaining each minute after the enzyme was added to the substrate. W shows the results obtained at a temperature of 40 °C.

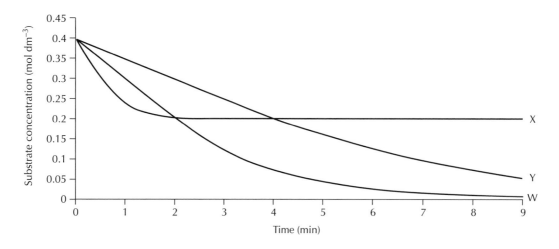

a) (i) Explain whether the temperature used for X was higher or lower than 40 °C.   [3]

(ii) Estimate the temperature that was used for Y.   [2]

b) Draw a curve on the graph to show the expected results of repeating the experiment at 40 °C with

(i) a fixed low concentration of non-competitive inhibitor.   [1]

(ii) a fixed low concentration of competitive inhibitor.   [2]

2 Reverse transcriptase is an enzyme found only in cells infected by certain viruses.

a)) Distinguish between the 3′ terminal and 5′ terminal in a chain of nucleotides.   [2]

b) Nucleic acids contain purines and pyrimidines. Compare purines and pyrimidines.   [3]

c) Distinguish between translation and transcription.   [5]

3 The diagram below represents the structure of lysozyme, a protein consisting of a single polypeptide, found in egg white.

a) State the name given to the shape of this type of protein.   [1]

b) State what is meant by the primary structure of a protein.   [1]

c) In the regions labelled X and Y two different types of secondary structure are found.

(i) Identify each type of secondary structure: [2]

(ii) State the type of bonding that is used to stabilize these structures.   [1]

d) Explain the importance of the tertiary structure of this protein to its function.   [2]

Glycolysis

INTRODUCING GLYCOLYSIS

Cell respiration involves the production of ATP using energy released by the oxidation of glucose, fat or other substrates. If glucose is the substrate, the first stage of cell respiration is a metabolic pathway called **glycolysis**. The pathway is catalysed by enzymes in the cytoplasm. Glucose is partially oxidized in the pathway and a small amount of ATP is produced. This partial oxidation is achieved without the use of oxygen, so glycolysis can form part of both aerobic and anaerobic respiration.

OXIDATION AND REDUCTION IN CELL RESPIRATION

Cell respiration involves many oxidation and reduction reactions. The figure (top right) compares the ways in which chemical substances can be oxidized and reduced. Hydrogen carriers accept hydrogen atoms removed from substrates in cell respiration. The most commonly used hydrogen carrier is NAD^+ (nicotinamide adenine dinucleotide). Hydrogen atoms consist of one proton and one electron. When two hydrogen atoms are removed from a respiratory substrate, NAD^+ accepts the electrons from both atoms and the proton from one of them.

$$NAD^+ + 2H \longrightarrow NADH + H^+$$

The figure (right) shows equations for some of the chemical changes that are part of cell respiration. It is possible to use the information in the figure (top right) to deduce whether each of them is an oxidation, a reduction or both.

CONVERTING GLUCOSE TO PYRUVATE IN GLYCOLYSIS

There are four main stages in glycolysis.

1. Two phosphate groups are added to a molecule of glucose to form hexose biphosphate. Adding a phosphate group is called **phosphorylation**. Two molecules of ATP provide the phosphate groups. The energy level of the hexose is raised by phosphorylation and this makes the subsequent reactions possible.

2. The hexose biphosphate is split to form two molecules of triose phosphate. Splitting molecules is called **lysis**.

3. Two atoms of hydrogen are removed from each triose phosphate molecule. This is an **oxidation**. The energy released by this oxidation is used to link on another phosphate group, producing a 3-carbon compound carrying two phosphate groups. NAD^+ is the hydrogen carrier that accepts the hydrogen atoms.

4. Pyruvate is formed by removing the two phosphate groups and by passing them to ADP. This results in **ATP formation**.

The figure (right) shows the main stages of glycolysis.

SUMMARY OF GLYCOLYSIS

- One glucose is converted into two pyruvates.
- Two ATP molecules are used per glucose but four are produced so there is a net yield of two ATP molecules.
- Two NAD^+ are converted into two $NADH + H^+$

Comparison of oxidation and reduction

Oxidation reactions	Reduction reactions
Addition of oxygen atoms to a substance.	Removal of oxygen atoms from a substance.
Removal of hydrogen atoms from a substance.	Addition of hydrogen atoms to a substance.
Loss of electrons from a substance.	Addition of electrons to a substance.

Examples of oxidations and reductions in cell respiration

$$Fe^{3+} + electron \longrightarrow Fe^{2+}$$

$$Fe^{2+} \longrightarrow Fe^{3+} + electron$$

$$succinate + FAD \longrightarrow fumarate + FADH_2$$

$$malate + NAD^+ \longrightarrow oxaloacetate + NADH + H^+$$

$$pyruvate + NADH + H^+ \longrightarrow lactate + NAD^+$$

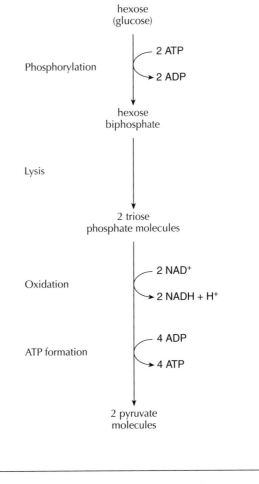

Stages of glycolysis

Krebs cycle

ANAEROBIC AND AEROBIC RESPIRATION

Glycolysis can occur without oxygen, so it forms the basis of anaerobic cell respiration. Pyruvate produced in glycolysis can only be oxidized further, with the release of more energy from it, if oxygen is available (right). This occurs in the mitochondrion. The first stage is called the **link reaction**. Enzymes in the matrix of the mitochondrion then catalyse a cycle of reactions called the **Krebs cycle**.

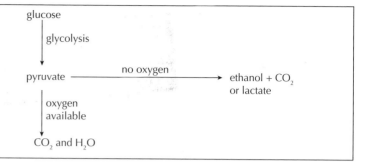

THE LINK REACTION

Pyruvate from glycolysis is absorbed by the mitochondrion. Enzymes in the matrix of the mitochondrion remove hydrogen and carbon dioxide from the pyruvate. The hydrogen is accepted by NAD^+. Removal of hydrogen is oxidation. Removal of carbon dioxide is decarboxylation. The whole conversion is therefore **oxidative decarboxylation**. The product of oxidative decarboxylation of pyruvate is an acetyl group, which is accepted by CoA (right).

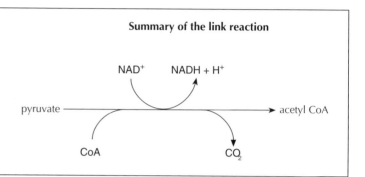

Summary of the link reaction

THE KREBS CYCLE

In the first reaction of the cycle an acetyl group is transferred from acetyl CoA to a four-carbon compound (oxaloacetate) to form a six-carbon compound (citrate). Citrate is converted back into oxaloacetate in the other reactions of the cycle. Three types of reaction are involved.

- Carbon dioxide is removed in two of the reactions. These reactions are **decarboxylations**. The carbon dioxide is a waste product and is excreted together with the carbon dioxide from the link reaction.
- Hydrogen is removed in four of the reactions. These reactions are **oxidations**. In three of the oxidations the hydrogen is accepted by NAD^+. In the other oxidation FAD accepts it. These oxidation reactions release energy, much of which is stored by the carriers when they accept hydrogen. This energy is later released by the electron transport chain and used to make ATP.
- ATP is produced directly in one of the reactions. This reaction is **substrate-level phosphorylation**.

The figure (right) is a summary of the Krebs cycle.

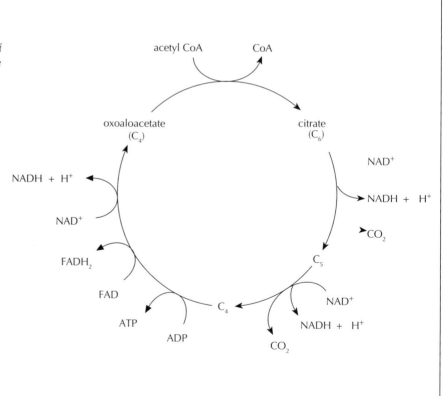

Summary of the Krebs cycle

Oxidative phosphorylation

THE ELECTRON TRANSPORT CHAIN

The electron transport chain is a series of electron carriers, located in the inner membrane of the mitochondrion. NADH supplies two electrons to the first carrier in the chain. The electrons come from oxidation reactions in earlier stages of cell respiration. The two electrons pass along the chain of carriers because they give up energy each time they pass from one carrier to the next. At three points along the chain enough energy is given up for ATP to be made by ATP synthase. As this ATP production relies on energy released by oxidation it is called **oxidative phosphorylation**. ATP synthase is also located in the inner mitochondrial membrane. $FADH_2$ also feeds electrons into the electron transport chain, but at a slightly later stage than NADH and at only two stages is sufficient energy released for ATP production by electrons from $FADH_2$.

THE ROLE OF OXYGEN

At the end of the electron transport chain the electrons are given to oxygen. At the same time oxygen accepts hydrogen ions, to form water. This happens in the matrix, on the surface of the inner membrane. This is the only stage at which oxygen is used in cell respiration. If oxygen is not available, electron flow along the electron transport chain stops and NADH + H⁺ cannot be reconverted to NAD⁺. Supplies of NAD⁺ in the mitochondrion run out and the link reaction and Krebs cycle cannot continue. Glycolysis can continue because conversion of pyruvate into lactate or ethanol and carbon dioxide produces as much NAD⁺ as is used in glycolysis. However, whereas aerobic cell respiration gives a yield of about 30 ATP molecules per glucose, glycolysis produces only two. Oxygen thus greatly increases the ATP yield.

The figure (below) shows the electron transport chain and the role of oxygen as the terminal electron acceptor.

The electron transport chain of mitochondria

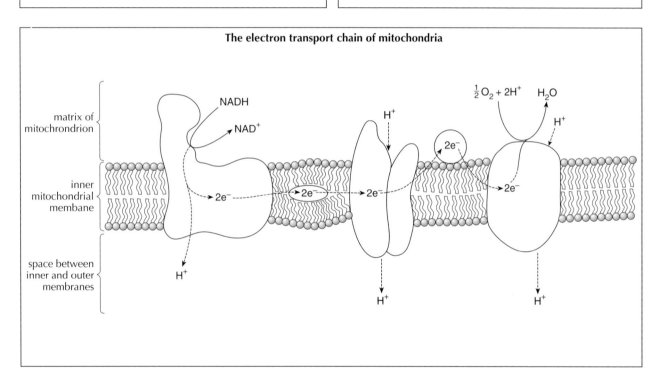

THE COUPLING OF ELECTRON TRANSPORT TO ATP SYNTHESIS

Energy released as electrons pass along the electron transport chain is used to pump protons (H⁺) across the inner mitochondrial membrane into the space between the inner and outer membranes. A concentration gradient is formed, which is a store of potential energy. ATP synthase, located in the inner mitochondrial membrane, transports the protons back across the membrane down the concentration gradient. As the protons pass across the membrane they release energy and this is used by ATP synthase to produce ATP. The coupling of ATP synthesis to electron transport via a concentration gradient of protons is called **chemiosmosis**.

The figure (right) shows some features of ATP synthase.

Structure of ATP synthase

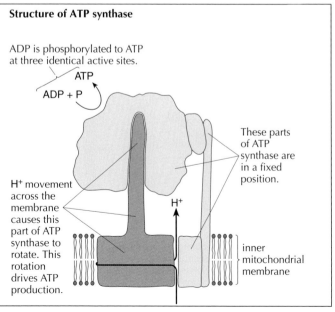

Mitochondria

The mitochondrion is an excellent example of the relationship between structure and function.
The figure (below) is an electron micrograph of a whole mitochondrion.
The figure (bottom) is a drawing of the same mitochondrion, labelled to show how it is adapted to carry out its function.

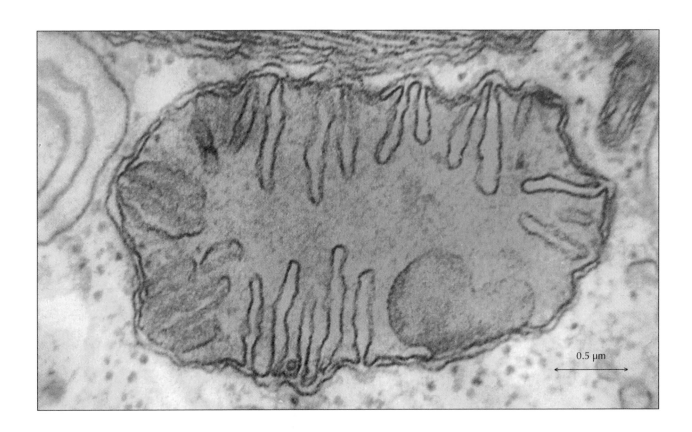

0.5 μm

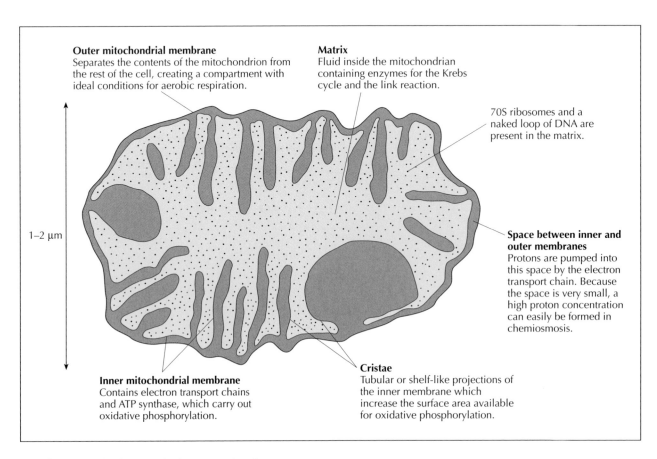

Outer mitochondrial membrane
Separates the contents of the mitochondrion from the rest of the cell, creating a compartment with ideal conditions for aerobic respiration.

Matrix
Fluid inside the mitochondrian containing enzymes for the Krebs cycle and the link reaction.

70S ribosomes and a naked loop of DNA are present in the matrix.

Space between inner and outer membranes
Protons are pumped into this space by the electron transport chain. Because the space is very small, a high proton concentration can easily be formed in chemiosmosis.

1–2 μm

Inner mitochondrial membrane
Contains electron transport chains and ATP synthase, which carry out oxidative phosphorylation.

Cristae
Tubular or shelf-like projections of the inner membrane which increase the surface area available for oxidative phosphorylation.

Light and photosynthesis

Photosynthesis is the process that plants, algae and some bacteria use to produce all of the organic compounds that they need. Photosynthesis involves many chemical reactions. Some of them need a continual supply of light and so are called **light-dependent reactions**. Other reactions need light indirectly, but can carry on for some time in darkness. These are called **light-independent reactions**. Glucose, amino acids and other organic compounds are produced in the light independent reactions. The light-dependent reactions produce intermediate compounds that are used in the light-independent reactions. In darkness these intermediate compounds are gradually used up.

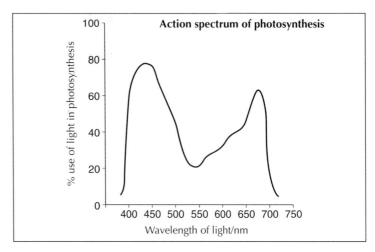

THE ACTION SPECTRUM OF PHOTOSYNTHESIS

A spectrum is a range of wavelengths of electromagnetic radiation. The spectrum of light is the range of wavelengths from 400 nm to 700 nm. Each wavelength is a pure colour of light:

    400–525 violet–blue
    525–625 green–yellow
    625–700 orange–red

The efficiency of photosynthesis is not the same in all wavelengths of light. The efficiency is the percentage of light of a wavelength that is used in photosynthesis. The figure (top right) is a graph showing the percentage use of the wavelengths of light in photosynthesis. This graph is called the **action spectrum** of photosynthesis. The graph shows that violet and blue light are used most efficiently and red light is also used efficiently. Green light is used much less efficiently.

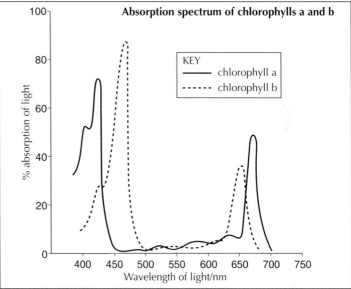

THE ABSORPTION SPECTRA OF PHOTOSYNTHETIC PIGMENTS

The action spectrum of photosynthesis is explained by considering the light-absorbing properties of the photosynthetic pigments. Most pigments absorb some wavelengths better than others. The figure (centre right) shows the percentage of the wavelengths of visible light that are absorbed by two common forms of chlorophyll. This graph is called the **absorption spectrum** of these pigments. The graph shows strong similarities with the action spectrum for photosynthesis.

- The greatest absorption is in the violet–blue range.
- There is a also a high level of absorption in the red range of the spectrum.
- There is least absorption in the yellow–green range of the spectrum. Most of this light is reflected.

There are some differences between the action spectrum and the absorption spectra. Whereas little light is absorbed by chlorophylls in the green to yellow range there is some photosynthesis. This is due to accessory pigments, including xanthophylls and carotene, which absorb wavelengths that chlorophyll cannot.

Action and absorption spectra of an alga

Some algae contain large amounts of accessory pigments. For example, kelp (*Laminaria saccharina*) contains carotene and fucoxanthin in addition to chlorophylls and so can absorb and use all wavelengths of light with about the same efficiency in photosynthesis. The graph below shows the action and absorption spectra for kelp. The colour of kelp can be deduced from the data.

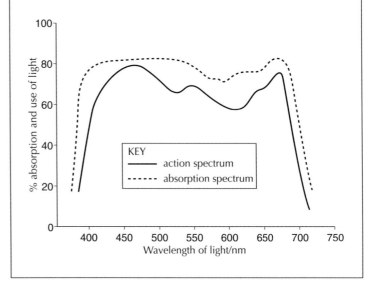

Light-dependent reactions

LIGHT ABSORPTION

Chlorophyll absorbs light and the energy from the light raises an electron in the chlorophyll molecule to a higher energy level. The electron at a higher energy level is an **excited electron** and the chlorophyll is **photoactivated**. In single chlorophyll molecules the excited electron soon drops back down to its original level, re-emitting the energy. Chlorophyll is located in thylakoid membranes and is arranged in groups of hundreds of molecules, called **photosystems**. There are two types of photosystem – photosystems I and II. Excited electrons from absorption of photons of light anywhere in the photosystem are passed from molecule to molecule until they reach a special chlorophyll molecule at the reaction centre of the photosystem. This chlorophyll passes the excited electron to a chain of electron carriers.

PRODUCTION OF ATP

An excited electron from the reaction centre of photosystem II is passed along a chain of carriers in the thylakoid membrane (below). It gives up some of its energy each time that it passes from one carrier to the next. At one stage, enough energy is released to make a molecule of ATP. The coupling of electron transport to ATP synthesis is by chemiosmosis, as in the mitochondrion. Electron flow causes a proton to be pumped across the thylakoid membrane into the fluid space inside the thylakoid. A proton gradient is created. ATP synthase, located in the thylakoid membranes, lets the protons across the membrane down the concentration gradient and uses the energy released to synthesize ATP.

The production of ATP using the energy from an excited electron from Photosystem II is called **non-cyclic photophosphorylation**. An alternative method of photophosphorylation is shown on page 81.

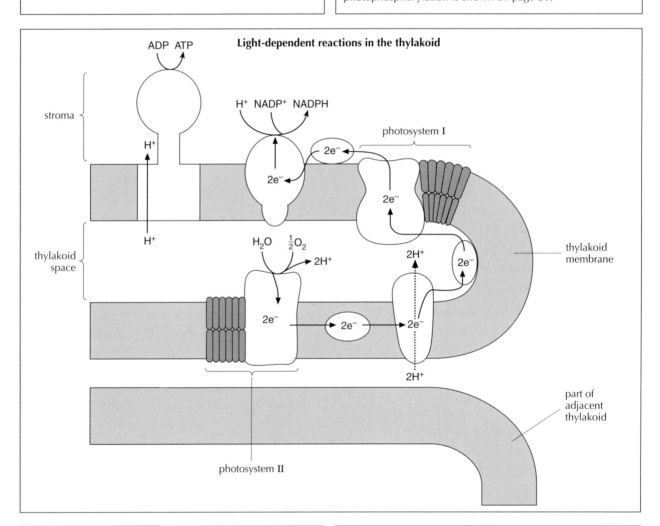

Light-dependent reactions in the thylakoid

PRODUCTION OF NADP

After releasing the energy needed to make ATP, the electron that was given away by photosystem II is accepted by photosystem I. The electron replaces one previously given away by photosystem I. With its electron replaced, photosystem I can be photoactivated by absorbing light and then give away another excited electron. This high-energy electron passes along a short chain of carriers to $NADP^+$ in the stroma. $NADP^+$ accepts two high-energy electrons from the electron transport chain and one H^+ ion from the stroma, to form NADPH.

PRODUCTION OF OXYGEN

Photosystem II needs to replace the excited electrons that it gives away. The special chlorophyll molecule at the reaction centre is positively charged after giving away an electron. With the help of an enzyme at the reaction centre, water molecules in the thylakoid space are split and electrons from them are given to chlorophyll. Oxygen and H^+ ions are formed as by-products. The splitting of water molecules only happens in the light, so is called **photolysis**. The oxygen produced in photosynthesis is all the result of photolysis of water. Oxygen is a waste product and is excreted.

Light-independent reactions

THE CALVIN CYCLE

The light-independent reactions take place within the stroma of the chloroplast. The first reaction involves a five-carbon sugar, ribulose bisphosphate (RuBP). RuBP is also a product of the light independent reactions, which therefore form a cycle, called the **Calvin cycle**. There are many alternative names for the intermediate compounds in the Calvin cycle. Glycerate 3-phosphate is sometimes also called 3-phosphoglycerate. Glycerate 3-phosphate is sometimes abbreviated as GP, which could be confused with glyceraldehyde 3-phosphate, which is a form of triose phosphate or with glucose phosphate. The abbreviation GP should therefore be avoided!

CARBON FIXATION

Carbon dioxide is an essential substrate in the light-independent reactions. It enters the chloroplast by diffusion. In the stroma of the chloroplast carbon dioxide combines with ribulose bisphosphate (RuBP), a five-carbon sugar, in a carboxylation reaction. The reaction is catalysed by the enzyme ribulose bisphosphate carboxylase, usually called **rubisco**. Large amounts of rubisco are present in the stroma, because it works rather slowly and the reaction that it catalyses is a very important one. The product of the reaction is a six-carbon compound, which immediately splits to form two molecules of glycerate 3-phosphate. This is therefore the first product of carbon fixation.

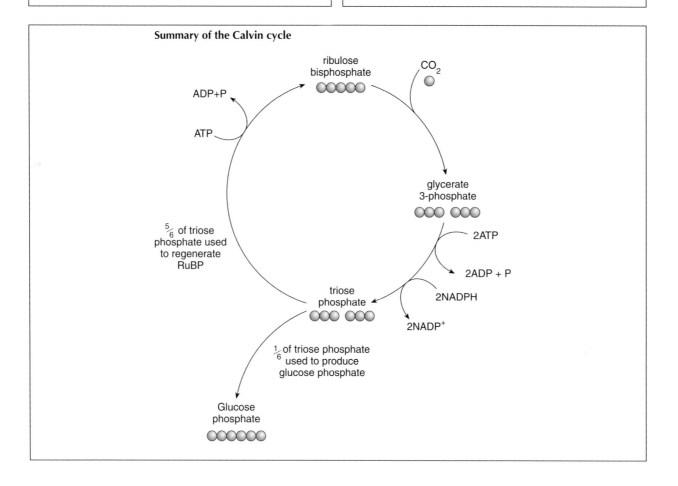

Summary of the Calvin cycle

REGENERATION OF RUBP

For carbon fixation to continue, one RuBP molecule must be produced to replace each one that is used. Triose phosphate is used to regenerate RuBP. Five molecules of triose phosphate are converted by a series of reactions into three molecules of RuBP. This process requires the use of energy in the form of ATP. The reactions can be summarized using equations where only the number of carbon atoms in each sugar molecule is shown.

$$C_3 + C_3 \longrightarrow C_6$$
$$C_6 + C_3 \longrightarrow C_4 + C_5$$
$$C_4 + C_3 \longrightarrow C_7$$
$$C_7 + C_3 \longrightarrow C_5 + C_5$$

For every six molecules of triose phosphate formed in the light-independent reactions, five must be converted to RuBP.

SYNTHESIS OF CARBOHYDRATE

Glycerate 3-phosphate, formed in the carbon fixation reaction, is an organic acid. It is converted into a carbohydrate by a reduction reaction. Hydrogen is needed to carry out this reaction and is supplied by NADPH. Energy is also needed and is supplied by ATP.
NADPH and ATP are produced in the light-dependent reactions of photosynthesis. Glycerate 3-phosphate is reduced to a three-carbon sugar, triose phosphate (TP). Linking together two triose phosphate molecules together produces glucose phosphate. Starch, the storage form of carbohydrate in plants, is formed in the stroma by condensation of many molecules of glucose phosphate.

Chloroplasts

The chloroplast is another example of close relationship between structure and function. The figure (below) is an electron micrograph of a chloroplast. The figure (bottom) is a drawing of the same chloroplast, labelled to show how it is adapted to carry out its function.

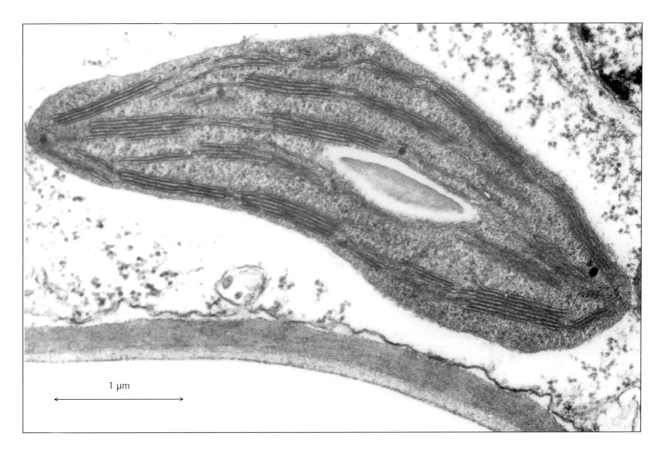

1 µm

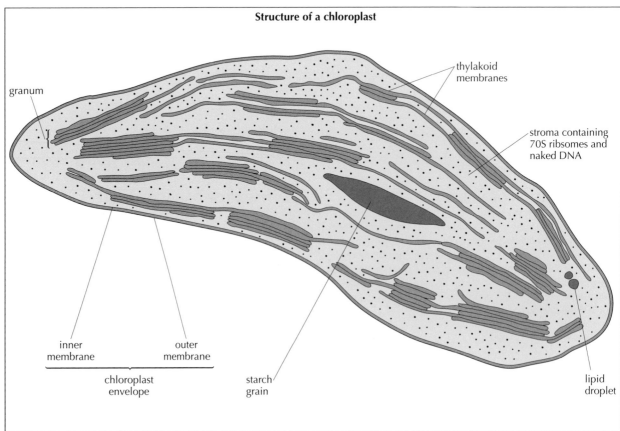

Structure of a chloroplast

granum

thylakoid membranes

stroma containing 70S ribsomes and naked DNA

inner membrane

outer membrane

chloroplast envelope

starch grain

lipid droplet

Limiting factors in photosynthesis

THE CONCEPT OF LIMITING FACTORS

Light intensity, carbon dioxide concentration and temperature are three factors that can determine the rate of photosynthesis. If the level of one of these factors is changed, the rate of photosynthesis changes. Usually, only changes to one of the factors will affect the rate of photosynthesis in a plant at a particular time. This is the factor that is nearest to its minimum and is called the **limiting factor**. Changing the limiting factor increases or decreases the rate, but changes to the other factors have no effect. This is because photosynthesis is a complex process involving many steps. The overall rate of photosynthesis in a plant is determined by the rate of whichever step is proceeding most slowly at a particular time. This is called the **rate-limiting step**. The three limiting factors affect different rate-limiting steps.

The figures on page 21 show the relationship between each of the limiting factors and the rate of photosynthesis.

THE EFFECT OF LIGHT INTENSITY

At low light intensities, there is a shortage of the products of the light-dependent reactions – NADPH and ATP. The rate-limiting step in the Calvin cycle is the point where glycerate 3-phosphate is reduced. At high light intensities some other factor is limiting.

Unless a plant is heavily shaded, or the sun is rising or setting, light intensity is not usually the limiting factor.

THE EFFECT OF CO_2 CONCENTRATION

At low and medium CO_2 concentrations, the rate-limiting step in the Calvin cycle is the point where CO_2 is fixed to produce glycerate 3-phosphate. RuBP and NADPH accumulate.

At high CO_2 concentrations some other factor is limiting.

Because the level of carbon dioxide in the atmosphere is never very high, carbon dioxide concentration is often the limiting factor.

THE EFFECT OF TEMPERATURE

At low temperatures, all of the enzymes that catalyse the reactions of the Calvin cycle work slowly. NADPH accumulates.

At intermediate temperatures, some other factor is limiting.

At high temperatures, RuBP carboxylase does not work effectively, so the rate-limiting step in the Calvin cycle is the point where CO_2 is fixed. NADPH accumulates.

Results of an investigation into limiting factors

The figure (right) shows the effects of light intensity on the rate of photosynthesis at two different temperatures and two carbon dioxide concentrations. It is possible to deduce which is the limiting factor at the point marked with an arrow (① – ④) on each curve.

KEY
——— 30 °C and 0.15% CO_2
- - - - 20 °C and 0.15% CO_2
× × × × 30 °C and 0.035% CO_2
∘ ∘ ∘ ∘ 20 °C and 0.035% CO_2

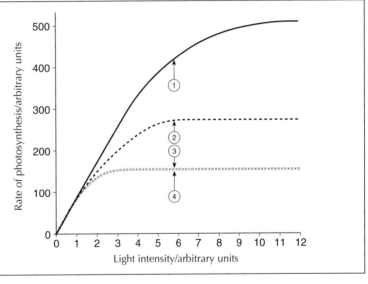

Summary of cyclic photophosphorylation

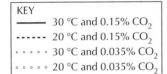

CYCLIC PHOTOPHOSPHORYLATION

When light is not the limiting factor, NADPH tends to accumulate in the stroma and there is a shortage of $NADP^+$. The normal flow of electrons in the thylakoid membranes is inhibited because $NADP^+$ is needed as a final acceptor of electrons. An alternative route can be used that allows ATP production when $NADP^+$ is not available. This pathway is called **cyclic photophosphorylation**.

- Photosystem I absorbs light and is photoactivated.
- Excited electrons are passed from photosystem I to a carrier in the chain between photosystem II and photosystem I.
- The electrons pass along the chain of carriers back to photosystem I.
- As the electrons flow along the chain of carriers they cause pumping of protons across the thylakoid membrane.
- A proton gradient is formed and this allows production of ATP by ATP synthase.

The figure (left) shows the pathway used in cyclic photophosphorylation.

EXAM QUESTIONS ON TOPIC 8

1 The electron micrograph below shows part of a plant root cell, including mitochondria.

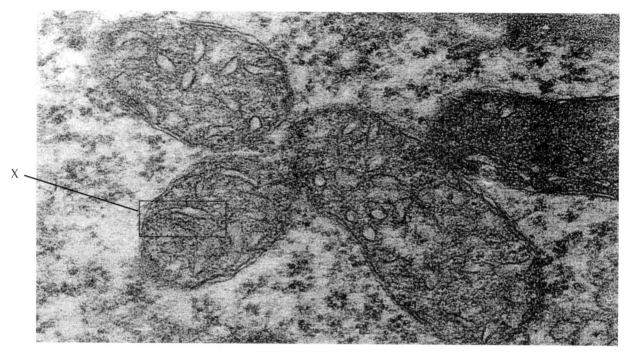

X

[Source: Dr B. E. Juniper, Dept. of Plant Sciences, University of Oxford]

    a) Explain briefly two features that allow the mitochondria in the micrograph to be identified. [2]

    b) Redraw the structure of the mitochondrion marked X. [2]

    c) Annotate the micrograph (not your drawing) to show one example of

        (i) a region where the Krebs cycle takes place

        (ii) a location of ATP synthetase

        (iii) a region where glycolysis takes place. [3]

2 a) Draw a curve of the action spectrum for photosynthesis on the axis below. [2]

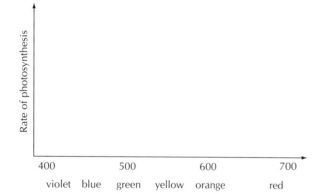

    b) Explain the relationship between the action spectrum and the absorption spectra of photosynthetic pigments. [3]

3 a) State two processes that involve chemiosmosis. [2]

    b) Explain the need for a membrane in chemiosmosis. [3]

    c) Suggest a location where chemiosmosis could occur in prokaryotes. [1]

Leaf structure and function

LEAVES AND PHOTOSYNTHESIS

The function of leaves is to produce food for the plant by photosynthesis. The leaf is adapted by its structure to carry out photosynthesis efficiently. On page 5 is a scanning electron micrograph of a leaf. The figure (below) is a plan diagram of tissues in part of a leaf of a dicotyledonous plant to show the adaptations for photosynthesis.

LEAVES AND TRANSPIRATION

Photosynthesis depends on gas exchange over a moist surface. Spongy mesophyll cell walls provide this surface. Water often evaporates from the surface and is lost, in a process called transpiration. *Transpiration is the loss of water vapour from the leaves and stems of plants.* The figure (below) shows adaptations to minimize the amount of transpiration.

Tissues of the leaf and their functions

Palisade mesophyll–consists of densely packed cylindrical cells with many chloroplasts. This is the main photosynthetic tissue and is positioned near the upper surface where the light intensity is highest

Upper epidermis–a continuous layer of cells covered by a thick waxy cuticle. Prevents water loss from the upper surface even when heated by sunlight. Lower epidermis in a cooler position has a thinner waxy cuticle

The main part of the leaf is the leaf blade or lamina. It has a large surface area to absorb sunlight but is very thin–only about 0.3 mm. It is composed of four thin tissue layers with veins at intervals.

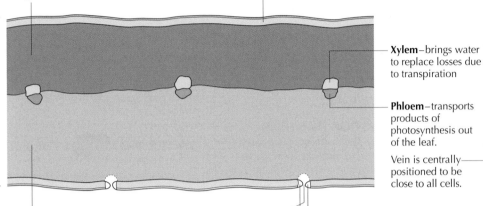

Xylem–brings water to replace losses due to transpiration

Phloem–transports products of photosynthesis out of the leaf.

Vein is centrally positioned to be close to all cells.

Spongy mesophyll–consists of loosely packed rounded cells with few chloroplasts. This tissue provides the main gas exchange surface so must be near the stomata in the lower epidermis.

Stoma–a pore that allows CO_2 for photosynthesis to diffuse in and O_2 to diffuse out.

Guard cells–this pair of cells can open or close the stoma and so control the amount of transpiration.

FACTORS AFFECTING TRANSPIRATION

Plants lose water vapour from their stems and leaves by transpiration. The rate of water loss varies depending on internal and external conditions. The main internal condition is whether the stomata are open or closed. The plant hormone **abscisic acid** causes guard cells to close the stomata. Plants produce abscisic acid when they are suffering water stress. External variables are called **abiotic factors** – four of these have an effect on the rate of transpiration.

- Light – guard cells close the stomata in darkness, so transpiration is much greater in the light.
- Temperature – heat is needed for evaporation of water from the surface of spongy mesophyll cells, so as temperature rises the rate of transpiration rises. Higher temperatures also increase the rate of diffusion through the air spaces in the spongy mesophyll, and reduce the relative humidity of the air outside the leaf.
- Humidity – water diffuses out of the leaf when there is a concentration gradient between the air spaces inside the leaf and the air outside. The air spaces are always nearly saturated. The lower the humidity outside the leaf, the steeper the gradient and therefore the faster the rate of transpiration.
- Wind – pockets of air saturated with water vapour tend to form near stomata in still air, which reduce the rate of transpiration. Wind blows the saturated air away and so increases the rate of transpiration.

TRANSPIRATION IN XEROPHYTES

Plants that are adapted to grow in very dry habitats are called **xerophytes**. *Cereus giganteus*, the saguaro or giant cactus, is an example of a xerophyte. It grows in deserts in Mexico and Arizona and shows many xerophytic adaptations, which help to reduce transpiration.

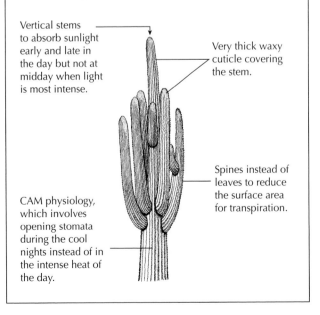

Vertical stems to absorb sunlight early and late in the day but not at midday when light is most intense.

Very thick waxy cuticle covering the stem.

Spines instead of leaves to reduce the surface area for transpiration.

CAM physiology, which involves opening stomata during the cool nights instead of in the intense heat of the day.

Transport and support

MINERAL UPTAKE BY ROOTS

Roots absorb water and mineral ions from the soil.
Plants increase the surface area for absorption by **branching of roots** and the growth of root hairs.

Plants absorb potassium, phosphate, nitrate and other mineral ions from the soil. The concentration of these ions in the soil is usually much lower than inside root cells, so they are absorbed by **active transport**. Root hair cells have mitochondria and protein pumps in their plasma membranes. Most roots only absorb mineral ions if they have a supply of oxygen, because they produce ATP for active transport, by aerobic cell respiration.

The rate of absorption of mineral ions is sometimes limited by the rate at which the ions move through the soil to the root. There are three ways in which ions can move:

- diffusion of mineral ions
- mass flow of water carrying ions, when water drains through the soil
- into fungal hyphae, that grow around plant roots in a mutualistic relationship, and then from the hyphae to the roots.

STRUCTURE AND FUNCTION OF STEMS

Stems connect the leaves, roots and flowers of plants and transport materials between them using xylem and phloem tissue. Stems support the aerial parts of terrestrial plants. Support is provided in several ways.

- Cells absorb water and high pressure develops inside the cell. This is **cell turgor** and it makes the cell almost rigid.
- Some cells develop thickened cellulose walls, which strengthen the plant.
- Cell walls in xylem tissue are both thickened and lignified making them very strong (above right). Xylem provides support especially in woody stems.

The figure (below) is a plan diagram to show the position of the tissues in the stem of a young dicotyledonous plant.

Transverse section of a stem

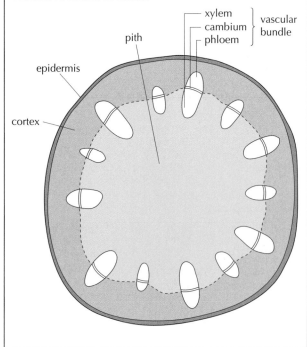

Structure of xylem vessels

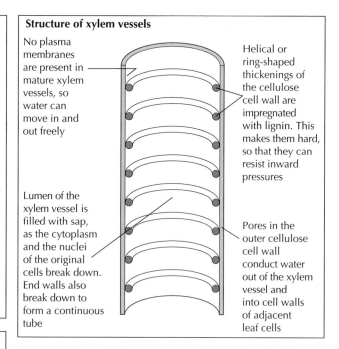

No plasma membranes are present in mature xylem vessels, so water can move in and out freely

Helical or ring-shaped thickenings of the cellulose cell wall are impregnated with lignin. This makes them hard, so that they can resist inward pressures

Lumen of the xylem vessel is filled with sap, as the cytoplasm and the nuclei of the original cells break down. End walls also break down to form a continuous tube

Pores in the outer cellulose cell wall conduct water out of the xylem vessel and into cell walls of adjacent leaf cells

WATER TRANSPORT THROUGH PLANTS

Xylem vessels contain long unbroken columns of water. When transpiration is occurring, water moves upwards from the roots to the leaves. This flow is called the **transpiration stream**.

The figure (above) shows the structure of a xylem vessel. Mature xylem vessels are dead and the flow of water through them is passive. Heat from the environment provides energy for **evaporation** of water from the cell walls of spongy mesophyll cells in the leaf. The water that evaporates is replaced with water from xylem vessels in the leaf. The water is pulled out of xylem vessels and through pores in spongy mesophyll cell walls by capillary action. Low pressure or suction is created inside xylem vessels when water is pulled out. This is called the **transpiration pull**. The suction extends down through the columns of water in xylem vessels to the roots. These columns of water do not usually break because of the **cohesion** of water molecules. Water molecules are cohesive due the hydrogen bonds between them.

Another process that can help water to move up in xylem vessels is the **adhesion** of water to the wall of the vessel. This is particularly important when sap starts to rise, in xylem vessels of plants that were leafless through the winter. In these plants, xylem vessels are empty in winter and refill in spring. Adhesion also helps prevent the column of water in water-filled xylem vessels from breaking.

TRANSPORT IN PHLOEM

Sugars and amino acids are transported inside plants by phloem tissue. This process is called **active translocation** because phloem cells have to use energy to make it happen. Sugars and amino acids are loaded into the phloem in parts of the plant called **sources** and are translocated to **sinks**, where they are unloaded. Examples of sources are parts of the plant where photosynthesis is occurring (stems and leaves) and storage organs where the stores are being mobilized. Examples of sinks are roots, growing fruits and the developing seeds inside them.

Reproduction of flowering plants

STRUCTURE AND FUNCTION OF FLOWERS

Flowers are the structures used by flowering plants for sexual reproduction. Female gametes are contained in ovules in the ovaries of the flower. Pollen grains, produced by the anthers, contain the male gametes. A zygote is formed by the fusion of a male gamete with a female gamete inside the ovule. This process is called **fertilization**.

Before fertilization, another process called **pollination** must occur. *Pollination is the transfer of pollen from an anther to a stigma.* Pollen grains containing male gametes cannot move without help from an external agent. Most plants use either wind or an animal for pollination. The structure of a flower is adapted to its method of pollination. The figure (below) shows the structure of a flower of *Lamium album*, which is adapted to bee pollination.

Pollen grains germinate on the stigma of the flower and a pollen tube containing the male gametes grows down the style to the ovary. The pollen tube delivers the male gametes to an ovule, which they fertilize.

Fertilized ovules develop into seeds. The figure (bottom) shows the structure of a seed of *Phaseolus multiflorus*. Ovaries containing fertilized ovules develop into fruits. The function of the fruit is **seed dispersal**.

Structure of *Lamium album* flower

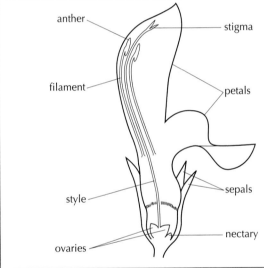

FACTORS NEEDED FOR SEED GERMINATION

Seeds will not germinate unless external conditions are suitable.

- Water must be available to rehydrate the dry tissues of the seed.
- Oxygen must be available for aerobic cell respiration. Some seeds respire anaerobically if oxygen is not available but ethanol produced in anaerobic respiration usually reaches toxic levels.
- Suitable temperatures are needed. Germination involves enzyme activity and at very low and very high temperatures enzyme activity is too slow. Some seeds remain dormant if temperatures are above or below particular levels, so that they only germinate during favourable times of the year.

The figure (below) shows the structure of a seedling of *Phaseolus multiflorus*, about 2 weeks after the start of germination.

Structure of a seedling of *Phaseolus multiflorus*

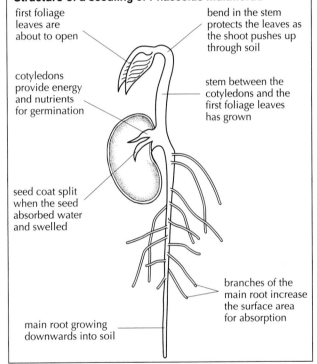

Structure of a seed of *Phaseolus multiflorus*

External structure

seed coat (testa)

scar where seed was attached to the ovary

Internal structure

embryo root (radicle)

embryo shoot (plumule)

cotyledon – one of two in the seed

seed coat

METABOLIC EVENTS DURING GERMINATION

- The first stage in germination is the absorption of water and the rehydration of living cells in the seed. This allows the cells to become metabolically active.
- Soon after absorbing water, a plant growth hormone called **gibberellin** is produced in the cotyledons of the seed.
- Gibberellin stimulates the production of amylase, which catalyses the digestion of starch into maltose in the food stores of the seed.
- Maltose is transported from the food stores to the growth regions of the seedling, including the embryo root and the embryo shoot.
- Maltose is converted into glucose, which is either used in aerobic cell respiration as a source of energy, or is used to synthesize cellulose or other substances needed for growth.

As soon as the leaves of the seedling have reached light and have opened, photosynthesis can supply the seedling with foods and the food stores of the seed are no longer needed.

Diversity in plant structure

MONOCOTYLEDONS AND DICOTYLEDONS

Flowering plants are divided into two groups, according to the number of leaves that the embryo plant has, inside the seed. Monocotyledons have one cotyledon (seed leaf) whereas dicotyledons have two. There are other differences:

Monocotyledons	Dicotyledons
Leaf veins run parallel to each other	Leaf veins form a net-like pattern
Vascular bundles are spread through the stem randomly	Vascular bundles are in a ring near the outside of the stem
Stamens and other organs in the flower are in multiples of 3	Stamens and other floral organs are in multiples of 4 or 5
Unbranched roots grow from stems. They are called adventitious roots	Roots branch off from other roots. Often there is a main tap root

Examples of plants in each group are shown in the figures (below).

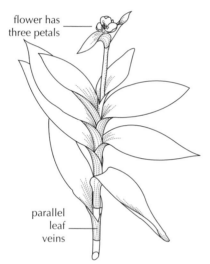

flower has three petals

parallel leaf veins

Tradescantia pallida

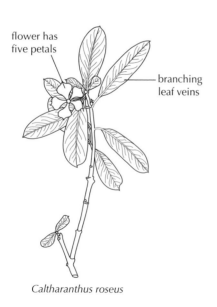

flower has five petals

branching leaf veins

Caltharanthus roseus

MODIFIED ROOTS, STEMS AND LEAVES

The normal functions of the roots, stems and leaves of plants have been described on the previous pages. In some plants, these organs have become modified for other functions.

1. Bulbs

In some monocotyledon plants, leaf bases grow to form an underground organ called a bulb (below).
Plants use bulbs for food storage. They can be identified from the series of leaf bases fitting inside each other, with a central shoot apical meristem.

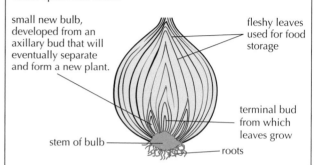

small new bulb, developed from an axillary bud that will eventually separate and form a new plant.

fleshy leaves used for food storage

stem of bulb

terminal bud from which leaves grow

roots

2. Stem tubers

In some dicotyledon plants, stems grow downwards into the soil and sections of them grow into stem tubers (below). They are used for food storage. They can be identified as stems because despite being swollen their vascular bundles are arranged in a ring.

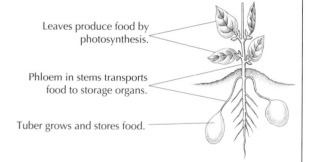

Leaves produce food by photosynthesis.

Phloem in stems transports food to storage organs.

Tuber grows and stores food.

3. Storage roots

Some roots become swollen with stores of food (below). They can easily be identified from their shape and from vascular tissue being in the centre.

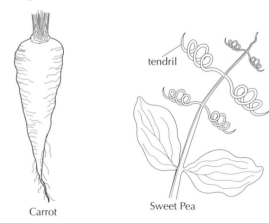

tendril

Carrot

Sweet Pea

4. Tendrils

Tendrils are narrow outgrowths from leaves that rotate through the air until they touch a solid support, to which they attach, allowing the plant to climb upwards (above).

Growth and development in plants

APICAL AND LATERAL MERISTEMS

Plants have regions where cells continue to divide and grow, often throughout the life of the plant. These regions are called **meristems**. Flowering plants all have meristems at the tip of the root and the tip of the stem (below). These are called **apical meristems** as they are at the apex of the root and stem. Growth in apical meristems allows roots and stems to elongate. The shoot apical meristem also produces new leaves and flowers.

Many dicotyledonous plants also develop **lateral meristems**. In young stems, this consists of cambium in the vascular bundles, but as the stem grows older, a complete ring of cambium forms.

A similar lateral meristem forms in older roots. Growth in lateral meristems makes roots and stems thicker, with extra xylem and phloem tissue. The growth in thickness of tree trunks is due to the lateral meristem, inside the bark.

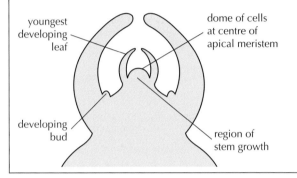

PHOTOPERIODIC CONTROL OF FLOWERING

Some plants only flower at the time of year when days are short and other plants only flower when the days are long. They are called short-day plants and long-day plants. Experiments have shown that it is not the length of day but the length of night that is significant. For example, chrysanthemums are short-day plants and only flower when they receive a long continuous period of darkness (below). They therefore naturally flower in the autumn (fall). Growers can produce pots of flowering chrysanthemums at all times of the year by keeping them in greenhouses with blinds. When the nights are not long enough to induce flowering, the blinds are closed to extend the nights artificially. In a similar way, petunias, which are long-day plants, can be induced to flower at times of the year when the days are short by being given extra light in greenhouses to reduce the length of the nights.

Response of chrysanthemums to different light/dark regimes

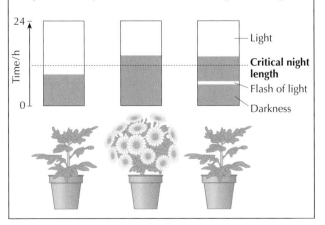

AUXIN AND PHOTOTROPISM

Plants use hormones to control their growth and their development. An example of a plant hormone is auxin, which acts as a growth promoter. It does this by causing secretion of hydrogen ions into cell walls, which loosens connections between cellulose fibres, allowing cell expansion.

One of the processes that auxin controls is phototropism – directional growth in response to the source of light. Shoot tips can detect the source of the brightest light. They also produce auxin. According to a long-standing theory, auxin is redistributed in the shoot tip from the lighter side to the shadier side. It then promotes more growth on the shadier side, causing the shoot to bend towards the light.

Molecular mechanisms for the action of auxin are being discovered. There are pumps in the plasma membrane called auxin efflux carriers. These are distributed unevenly and so can redistribute auxin in a tissue. Plant cells contain an auxin receptor. When auxin binds to it, transcription of specific genes is promoted, which affect the growth of the cell in the ways described above.

PHYTOCHROME AND PHOTOPERIODISM

Plants can measure the length of periods of dark to an accuracy of a few minutes. They do this using a pigment in their leaves called **phytochrome**, which exists in two interconvertible forms. One form is called P_r because it absorbs red light with a wavelength of 660 nm. P_r is the inactive form of phytochrome. When it absorbs red light it is rapidly converted into the active form, called P_{fr}. This form can absorb far red light with a wavelength of 730 nm and is then rapidly converted back to P_r. In normal daylight there is much more red light than far red light so phytochrome exists in the active P_{fr} form. In darkness P_{fr} reverts very slowly to P_r (above). This gradual reversion process is probably how the length of the dark period is timed. Enough P_{fr} remains in long-day plants at the end of short nights to stimulate flowering. In *Arabidopsis*, which is a long-day plant, a protein has been found to which P_{fr} binds. This protein probably acts as a transcription factor, causing genes involved in flowering to be switched on. In short day plants P_{fr} presumably acts as an inhibitor of flowering. At the end of long nights, enough P_{fr} has been converted to P_r to allow flowering to occur.

Interconversions of phytochrome

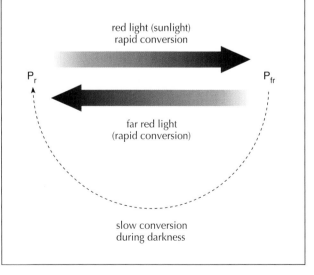

1 Control of flowering in long-day and short-day plants involves inter-conversion of phytochrome between its two forms, P_r and P_{fr}.

   a) State whether phytochrome is in the P_r or P_{fr} form at the end of the day (sunset) in

   (i) long-day plants   [1]

   (ii) short-day plants   [1]

   b) Explain how long-day and short-day plants time the length of the night.   [2]

   c) Distinguish between the effect of P_{fr} in long-day and short-day plants.   [2]

2 Flowering plants (angiospermophytes) are classified into two groups: monocotyledons and dicotyledons.

   a) Outline three differences between monocotyledons and dicotyledons.   [3]

   b) Distinguish between growth due to apical and lateral meristems in the stems of dicotyledons. [2]

   c) Monocotyledons do not have lateral meristems. Predict the consequences for monocotyledons of not having lateral meristems.   [2]

3 C_3 and CAM plants both need CO_2 for photosynthesis. They take in CO_2 through microscopic pores called stomata. The stomata can very from being fully closed (0% open) to fully open (100% open). The circular graph below shows the width of opening of stomata during a 24 hour period in a C_3 plant and a CAM plant.

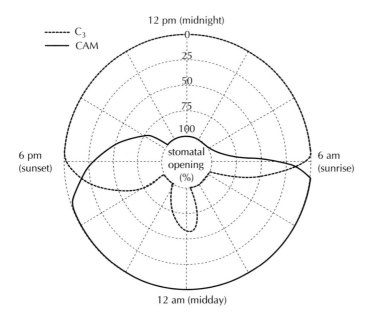

   a) Identify the hours during which stomata were fully closed in

   (i) the C_3 plant   [1]

   (ii) the CAM plant   [1]

   b) One of the two plants is a xerophyte.
   Use the data in the graph to predict whether the C_3 plant or the CAM plant is the xerophyte. [2]

   c) (i) Outline the changes in the stomata of the C_3 plant shown in the graph between 11.00 am. and 2.00 pm. [2]

   (ii) Suggest a reason for the changes.   [1]

Mendel's law of independent assortment

Gregor Mendel discovered the Law of Segregation by doing monohybrid crosses with pea plants. He discovered another law of inheritance by doing crosses in which the parents differed in two characteristics, that are controlled by two different genes. These are called **dihybrid crosses**. Mendel did his dihybrid crosses with pea plants. An example of one of his crosses is shown below. The parents in this cross differ in seed shape, controlled by one gene, and in seed colour, controlled by a different gene.

KEY TO SYMBOLS

S = allele for smooth seed.
s = allele for wrinkled seed.
Y = allele for yellow seed.
y = allele for green seed.

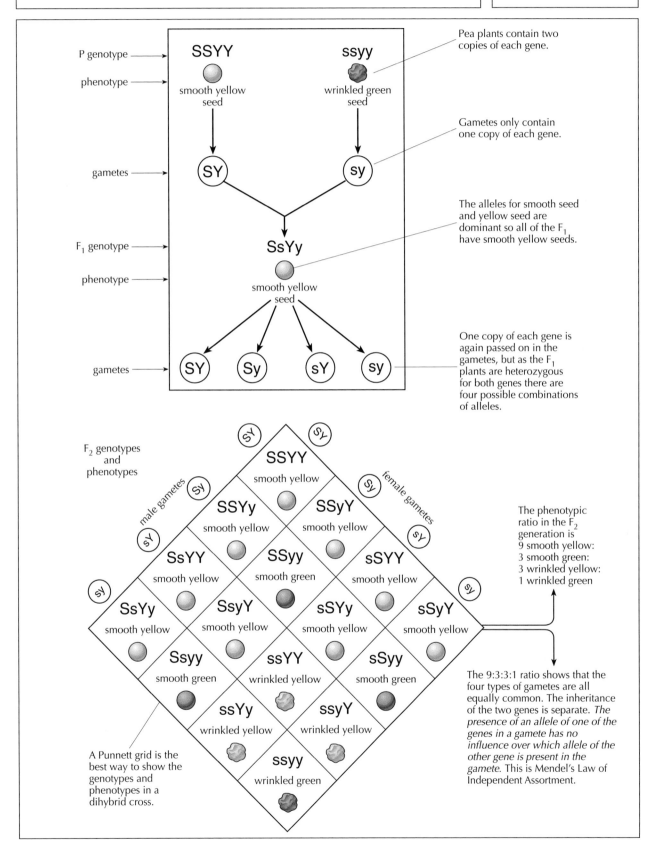

Pea plants contain two copies of each gene.

Gametes only contain one copy of each gene.

The alleles for smooth seed and yellow seed are dominant so all of the F_1 have smooth yellow seeds.

One copy of each gene is again passed on in the gametes, but as the F_1 plants are heterozygous for both genes there are four possible combinations of alleles.

The phenotypic ratio in the F_2 generation is
9 smooth yellow:
3 smooth green:
3 wrinkled yellow:
1 wrinkled green

The 9:3:3:1 ratio shows that the four types of gametes are all equally common. The inheritance of the two genes is separate. *The presence of an allele of one of the genes in a gamete has no influence over which allele of the other gene is present in the gamete.* This is Mendel's Law of Independent Assortment.

A Punnett grid is the best way to show the genotypes and phenotypes in a dihybrid cross.

Dihybrid crosses

PREDICTING RATIOS IN DIHYBRID RATIOS

The 9:3:3:1 ratio is often found when parents that are heterozygous for two genes are crossed together. The ratio is the product of two 3:1 ratios – each of the two genes would give a 3:1 ratio in a monohybrid cross between two heterozygous parents. In a dihybrid cross they follow Mendel's Law of Independent Assortment because they are unlinked.

Dihybrid crosses can give other ratios if:
- either of the genes has codominant alleles,
- either of the parents is homozygous for one/both of the genes,
- either of the genes is sex linked. Sex-linked genes are located on sex chromosomes instead of on autosomes (non-sex chromosomes).

The figure (right) shows ratios that these types of genes could give. Another cause of unusual ratios is interaction between genes. The figure (below) shows an example of a dihybrid cross where there is interaction between genes.

Possible ratios in dihybrid crosses

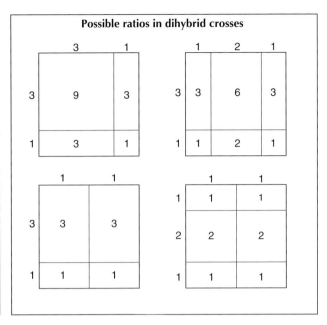

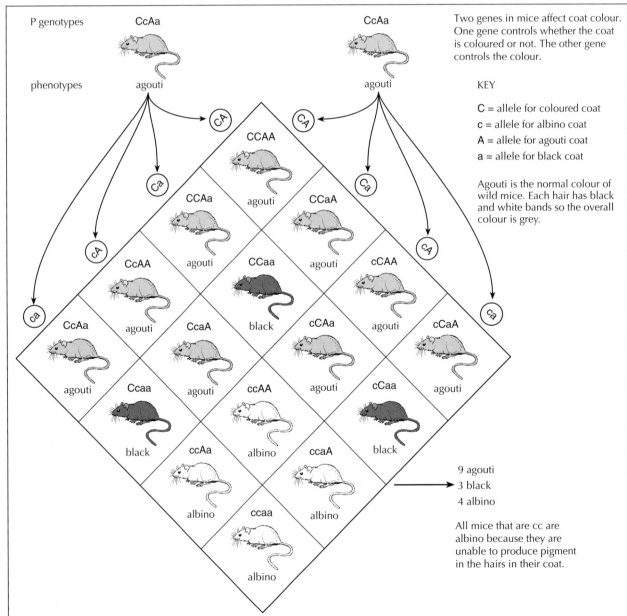

P genotypes — CcAa — CcAa

phenotypes — agouti — agouti

Two genes in mice affect coat colour. One gene controls whether the coat is coloured or not. The other gene controls the colour.

KEY

C = allele for coloured coat
c = allele for albino coat
A = allele for agouti coat
a = allele for black coat

Agouti is the normal colour of wild mice. Each hair has black and white bands so the overall colour is grey.

CCAA agouti
CCAa agouti
CCaA agouti
CcAA agouti
CCaa black
cCAA agouti
ccaa agouti → 9 agouti, 3 black, 4 albino...

CcAa agouti
CcaA agouti
CCaa black
cCAa agouti
cCaA agouti

Ccaa black
ccAA albino
ccaA albino
cCaa black

ccAa albino
ccaA albino

ccaa albino

9 agouti
3 black
4 albino

All mice that are cc are albino because they are unable to produce pigment in the hairs in their coat.

Polygenic inheritance

THE DISCOVERY OF POLYGENIC INHERITANCE

Some characteristics are influenced by more than one gene. This is called polygenic inheritance. Gregor Mendel discovered an example of **polygenic inheritance**, when he crossed a purple-flowered species of bean with a white-flowered species. The F_1 offspring were all purple, so he expected a 3:1 ratio of purple to white flowers in the F_2 offspring. Instead, he found a much smaller proportion of white flowers and a wide variety of shades of purple flower. Mendel suggested that two or three genes might be involved. If these were codominant genes, each with two alleles, one for purple flowers (A^P and B^P) and one for white (A^W and B^W), there could be five shades of flower colour (right).

POLYGENIC INHERITANCE AND CONTINUOUS VARIATION

Most examples of polygenic inheritance involve more than two genes with codominant alleles. As the number of genes involved increases, the number of possible phenotypes increases. Eventually, it becomes impossible to divide individuals into discrete groups – the variation is continuous.

Results of a cross between red and white flowered beans

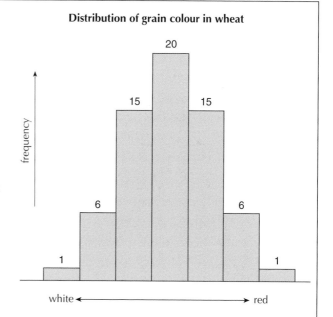

EXAMPLES OF POLYGENIC INHERITANCE
Grain colour in wheat

Wheat grains vary in colour from white to dark red, depending on the amount of a red pigment they contain. Three genes control the colour. Each gene has two alleles, one that causes pigment production and one that does not. Wheat grains can therefore have between 0 and 6 alleles for pigment production. The figure (right) shows the expected distribution of grain colour from a cross between two plants that are heterozygous for each of the three genes.

Skin colour in humans

The colour of human skin depends on the amount of the black pigment melanin in it. There is a continuous distribution of skin colour from very pale (little melanin) to black (much melanin). At least four and possibly more genes are involved, each with alleles that promote melanin production and alleles that do not. There is therefore a wide range of possible genotypes with anything from no alleles promoting melanin production to many.
The figure (below) shows humans with a range of skin colour.

Distribution of grain colour in wheat

Skin colour variation in humans

Genes – linked and unlinked

UNLINKED GENES

Mendel's law of independent assortment can be explained in terms of chromosome movements during meiosis. If pairs of genes are located on different types of chromosome, when homologous chromosomes pair up in meiosis the genes are on different pairs. The pairs of homologous chromosomes are called **bivalents**. The bivalents are orientated randomly on the equator, so the pole to which alleles on other bivalents are moving does not affect the pole to which alleles on a bivalent move. Random orientation of bivalents allows combinations of alleles to be broken up, so that new combinations can be formed when gametes fuse during fertilization.

If two parents with the genotypes AABB and aabb are crossed together, the gametes that they produce (AB and ab) will fuse together to give an F_1 hybrid with the genotype AaBb. The figure below shows the possible gametes that could be produced by meiosis in this F_1 hybrid. The parents could not have produced two of the gametes (Ab and aB).

GENE LINKAGE

Some pairs of genes do not follow the law of independent assortment. The expected 9:3:3:1 ratio is not found when parents that are heterozygous for the two genes are crossed. The figure (below) shows the first example of this to be discovered. The results show that there were more offspring than expected with the parental character combinations – purple long and red round. There were fewer than expected with the new combinations – purple round and red long. Combinations of genes tend to be inherited together. This is called **gene linkage**.

Gene linkage is caused by pairs of genes being located on the same type of chromosome. New combinations of alleles can only be produced if DNA is swapped between chromatids. This is called **recombination** and involves a special process called **crossing over**, which happens during the early stages of meiosis. Individuals that have a different combination of characters from parents, as a result of crossing over, are called **recombinants**.

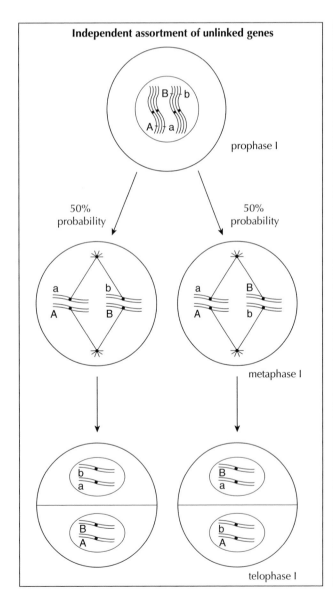

Independent assortment of unlinked genes

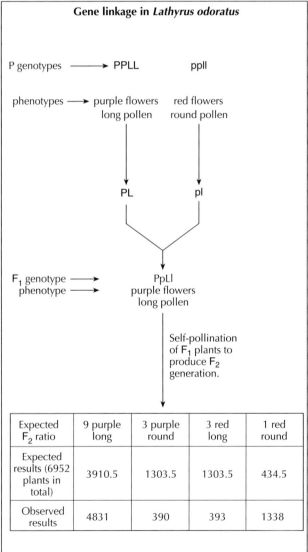

Gene linkage in *Lathyrus odoratus*

Expected F_2 ratio	9 purple long	3 purple round	3 red long	1 red round
Expected results (6952 plants in total)	3910.5	1303.5	1303.5	434.5
Observed results	4831	390	393	1338

Crossing-over

EVENTS IN PROPHASE I OF MEIOSIS

Homologous chromosomes pair up in prophase I of meiosis. Each homologous chromosome consists of two sister chromatids. Chromatids of different chromosomes are called non-sister chromatids. While the chromosomes are paired, sections of chromatid are exchanged in a process called **crossing-over**.
The figure (right) shows how crossing-over occurs.

BENEFITS OF CROSSING-OVER

Crossing-over has two important consequences.
1. It creates chiasmata which hold homologous chromosomes together in pairs called bivalents, during the later stages of prophase I and metaphase I until microtubules have attached.
2. It allows recombination of linked genes. All of the genes that have their loci on the same chromosome type form a **linkage group**.
Recombination of genes in a linkage group cannot occur without crossing-over. The point where crossing-over occurs along chromosomes is random – it can occur at a vast number of different points. Meiosis can therefore produce an almost infinite amount of genetic variety.
The figure (below) shows how crossing-over can cause recombination of linked genes.
The figure (right) shows an example of a cross involving gene linkage, using bars to represent the chromosomes on which the genes are linked. A test cross was done on the F$_1$ plants.

Recombination of linked genes

Parental gene combinations are AB and ab

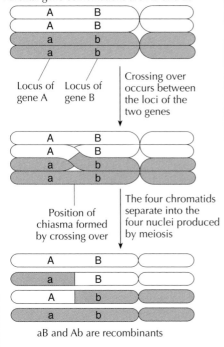

aB and Ab are recombinants

The process of crossing over

At one stage in prophase I all of the chromatids of two homologous chromosomes become tightly paired up together. This is called *synapsis*.

four chromatids in total, long and thin at this stage — centromeres

The DNA molecule of one of the chromatids is cut. A second cut is made at exactly the same point in the DNA of a non-sister chromatid.

DNA is cut at the same point in two non-sister chromatids

The DNA of each chromatid is joined up to the DNA of the non-sister chromatid. This has the effect of swapping sections of DNA between the chromatids.

In the later stages of prophase I the tight pairing of the homologous chromosomes ends, but the sister chromatids remain tightly connected. When each cross-over has occurred there is an X-shaped structure called a chiasma.

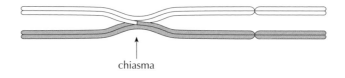

chiasma

AN EXAMPLE OF GENE LINKAGE AND TEST-CROSSING IN *ZEA MAYS*

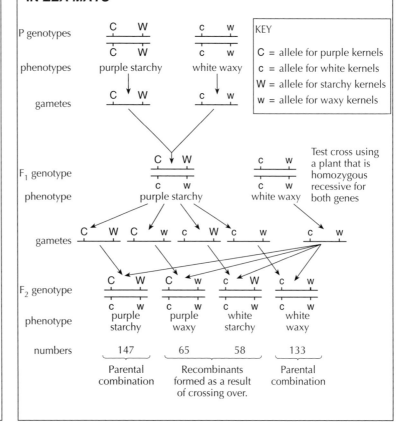

KEY
C = allele for purple kernels
c = allele for white kernels
W = allele for starchy kernels
w = allele for waxy kernels

Phases of meiosis

Meiosis involves two divisions. Each division is divided into four phases. The main events of each phase are listed below.

PROPHASE I
- Chromosomes start to coil up and so become shorter and thicker.
- Homologous chromosomes pair up.
- Crossing over occurs.
- Centrioles move to the poles in animal cells.
- Nucleoli break down.
- At the end of prophase I the nuclear membrane breaks down.

METAPHASE I
- Chromosomes continue to shorten and thicken.
- Spindle microtubules attach to the centromeres.
- Bivalents line up on the equator.
- Chiasmata slide towards the ends of the chromosomes, causing the shapes of the bivalents to change.
- At the end of metaphase I the chromosomes start to move.

ANAPHASE I
- The two chromosomes of each bivalent move to opposite poles. This halves the chromosome number. Each chromosome consists of two chromatids. Because of crossing over the two chromatids are not identical.
- At the end of anaphase I the chromosomes reach the poles.

TELOPHASE I
- Nuclear membranes form around the groups of chromosomes at each pole.
- The cell divides to form two haploid cells.
- The chromosomes uncoil partially.
- At the end of telophase I the two cells either enter a brief period of interphase or immediately proceed to the second division of meiosis. The DNA is not replicated.

PROPHASE II
- Chromosomes become shorter and thicken again by coiling.
- Centrioles move to the poles in animal cells.
- At the end of prophase II the nuclear membranes break down.

METAPHASE II
- Spindle microtubules attach to the centromeres.
- Chromosomes line up on the equator
- At the end of metaphase II the centromeres divide.

ANAPHASE II
- The two chromatids of each chromosome move to opposite poles.
- At the end of anaphase II the chromatids reach the poles.

TELOPHASE II
- Nuclear membranes form around the groups of chromatids at each pole. Each chromatid is now considered to be a chromosome.
- The two cells each divide to form to four cells in total.
- The chromosomes uncoil.
- Nucleoli appear.
- In most organisms the cells formed at the end of telophase II develop into gametes.

SUMMARY OF MEIOSIS
1. Meiosis involves two divisions. One cell or nucleus divides to form four cells or nuclei.
2. The chromosome number is halved, from diploid to haploid.
3. An almost infinite amount of genetic variety is produced, as a result of crossing-over in prophase I and the random orientation of bivalents in metaphase I.

The figure (below) shows micrographs of four stages in meiosis in cells from the testis of a locust.

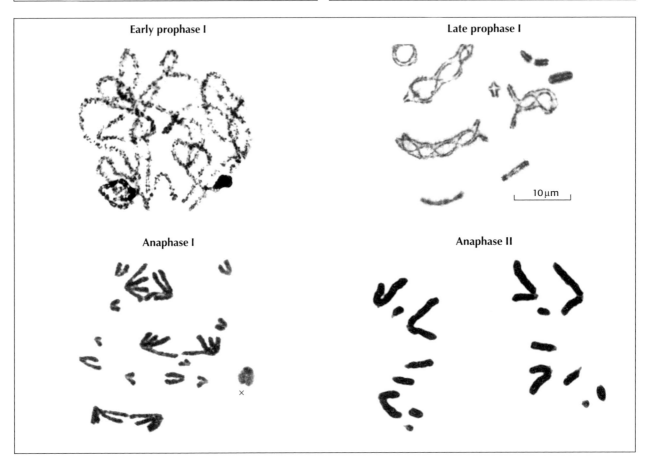

Early prophase I

Late prophase I

10 μm

Anaphase I

Anaphase II

EXAM QUESTIONS ON TOPIC 10

1 In some plants two genes control flower colour. [Note: – represents any allele]

Plants with the genotype A_B_ have blue flowers.

Plants with the genotype A_bb have red flowers.

Plants with the genotype aa _ _ have white flowers.

a) State the name given to the type of inheritance where more than one gene controls a single phenotypic characteristic. [1]

A homozygous blue-flowered plant (AABB) is crossed with a homozygous white-flowered plant (aabb).

b) State the genotype and phenotype of the F_1 offspring. [2]

c) The F_1 plants are allowed to pollinate each other. Deduce, using the Punnett grid below, the genotypes of the gametes produced by the F_1 plants and the genotypes and phenotypes of all the possible F_2 offspring. [5]

gametes →				

d) State the expected ratio of flower colours in the F_2 offspring. [1]

e) The two genes code for enzymes used to convert a white substance into a red pigment and the red pigment into a blue pigment. Deduce the effect of the enzymes produced from gene A and gene B. [1]

2 a) Define recombination. [1]

When grey bodied long winged Drosophila flies were test crossed with black bodied vestigial wing flies the F_1 generation was found to contain:

407 grey bodied long winged flies

396 black bodied vestigial winged flies

75 black bodied long winged flies

69 grey bodied vestigial winged flies

b) Identify which of the flies were recombinants. [2]

c) The F_1 generation does not follow Mendel's Second Law (Law of Independent Assortment). Explain how the observed ratio could have arisen. [3]

d) Suggest how geneticists could make use of experimental results of the type shown above. [2]

3 The micrograph below shows a pair of homologous chromosomes in a cell carrying out meiosis in the grasshopper (*Chorthippus parallelus*).

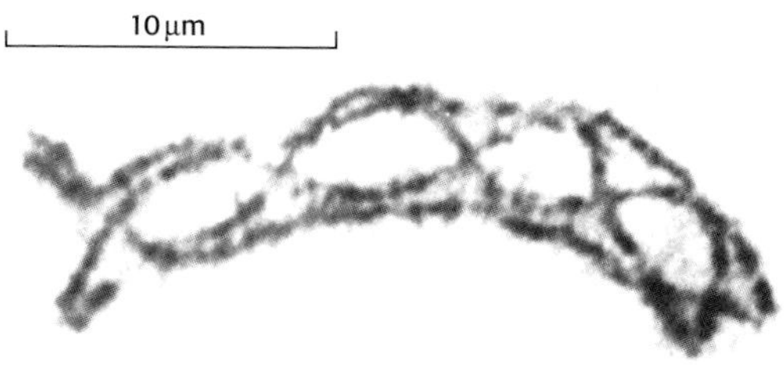

a) Identify the stage of meiosis of the cell that contained the pair of chromosomes. [2]

b) In the pair of chromosomes in the micrograph deduce the number of

(i) chromatids [1]

(ii) chiasmata [1]

c) Outline how chiasmata are produced. [3]

Antibody production

STAGES IN ANTIBODY PRODUCTION

The production of antibodies by the immune system is one of the most remarkable biological processes. When a pathogen invades the body, the immune system gears up to produce large amounts of the specific antibodies needed to combat the pathogen. This process only takes a few days. The production of antibodies by B-cells is shown in a simplified form on page 50. Antibody production usually depends on other types of lymphocyte, including macrophages and helper T-cells. The roles of these cells are explained here.

1. Antigen presentation

Macrophages take in antigens by endocytosis, process them and then attach them to membrane proteins called MHC proteins. The MHC proteins carrying the antigens are then moved to the plasma membrane by exocytosis and the antigens are displayed on the surface of the macrophage. This is antigen presentation.

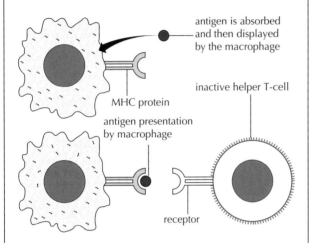

2. Activation of helper T-cells

Helper T-cells have receptors in their plasma membrane that can bind to antigens presented by macrophages. Each helper T-cell has receptors with the same antigen-binding domain as an antibody. These receptors allow a helper T-cell to recognize an antigen presented by a macrophage and bind to the macrophage. The macrophage passes a signal to the helper T-cell changing it from an inactive to an active state. This is activation of helper T-cells.

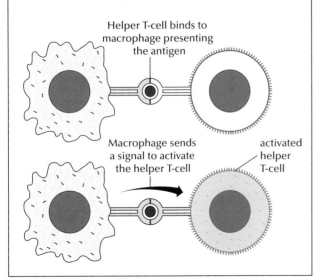

3. Activation of B-cells

Inactive B-cells have antibodies in their plasma membrane. If these antibodies match an antigen, the antigen binds to the antibody. An activated helper T-cell with receptors for the same antigen can then bind to the B-cell. The activated helper T-cell sends a signal to the B-cell, causing it to change from an inactive to an active state. This is activation of B-cells.

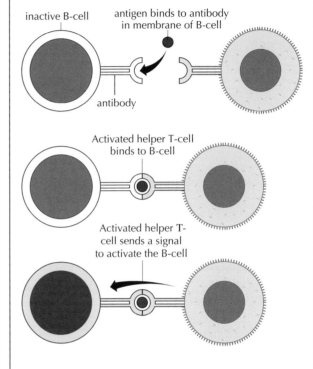

4. Production of plasma cells

Activated B-cells start to divide by mitosis to form a clone of cells. These cells become active, with a much greater volume of cytoplasm. They are then known as plasma cells. They have a very extensive network of rough endoplasmic reticulum. This is used for synthesis of large amounts of antibody, which is then secreted by exocytosis.

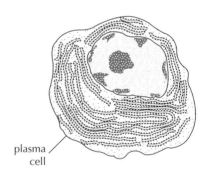

plasma cell

5. Production of memory cells

Memory cells are B-cells and T-cells that are formed at the same time as activated helper T-cells and B-cells, when a disease challenges the immune system. After the activated cells and the antibodies produced to fight the disease have disappeared, the memory cells persist and allow a rapid response if the disease is encountered again. Memory cells give long-term immunity to a disease.

Immunity and vaccination

ACTIVE AND PASSIVE IMMUNITY

Resistance to infection is called immunity. Antibodies give immunity to disease – this is sometimes called specific immunity, because one type of antibody gives protection against only one disease. Immunity can either be active or passive.

Active immunity is due to production of antibodies by the organism itself after the body's defence mechanisms have been stimulated by antigens.

An example is when infection with rubella virus causes immunity to rubella to develop and re-infection is very rare.

Passive immunity is due to the acquisition of antibodies received from another organism, in which active immunity has been stimulated.

Examples of passive immunity:
- During pregnancy, antibodies are passed across the placenta from mother to the fetus.
- The first milk produced after birth, called colostrum, contains antibodies that line the gut of newborn babies, helping to prevent infection.
- Antibodies are sometimes injected as an emergency treatment for virulent diseases, such as rabies.

PRINCIPLES OF ANTIBODY PRODUCTION

The immune system has the potential to produce a vast range of different types of antibody – perhaps 10^{15} different types. It would be impossible to make large quantities of all of these antibodies. Instead, a few B-cells that can make a type of antibody are produced and if these cells encounter an antigen to which their antibody binds, they multiply to form a clone of many cells. This is called **clonal selection**.

Sometimes several different types of antibody can bind to the same antigen, so more than one clone of cells is formed. This is called polyclonal selection.

A clone of B-cells can produce large amounts of antibody quickly and so give immunity to the disease with which the antigen is associated. Immunity to a disease is only developed if the disease challenges the immune system. This is called the principle of **challenge and response**.

These two principles do not fully explain antibody diversity. Research is ongoing into two additional processes:
- how lymphocytes splice together DNA taken from various parts of the genome, to produce a huge variety of genes coding for antibodies
- how rapid mutation occurs in antibody genes in lymphocytes that have been activated by antigen binding – this gives the chance of producing antibodies that fit the antigen better.

VACCINATION

A vaccine is a modified form of a disease-causing microorganism that stimulates the body to develop immunity to the disease, without fully developing the disease. Vaccines contain weakened forms of the microorganisms, killed forms or chemicals produced by the microorganism that act as antigens. The vaccine is either injected into the body or sometimes swallowed. The principle of vaccination is that antigens in the vaccine cause the production of the antibodies needed to control the disease. Sometimes two or more vaccinations are needed to stimulate the production of enough antibodies. The figure (right) shows a typical response to a first and second vaccination against a disease. The first vaccination causes a little antibody production and the production of some memory cells. The second vaccination, sometimes called a booster shot, causes a response from the memory cells and therefore faster and greater production of antibodies. Memory cells should persist to give long-term immunity.

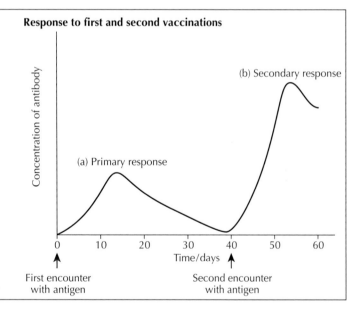

Response to first and second vaccinations

VACCINATION – BENEFITS AND DANGERS

There are huge benefits from vaccination:
1. Epidemics and pandemics can be prevented and some diseases can be completely eradicated – smallpox was the first and polio may be the second.
2. Deaths due to disease can be prevented. For example, measles is a major cause of death of unvaccinated children in some parts of the world.
3. Disability due to disease can be prevented, decreasing health care costs, for example deafness and blindness in babies whose mothers were not vaccinated and so contracted rubella during pregnancy.

Immunization is the most effective of all public health interventions. All vaccines are very carefully tested before being introduced and the risks of the diseases that they prevent are much greater than any adverse effects associated with vaccines themselves.

Unfortunately, the diseases that vaccines prevent become rare so that parents worry more about the vaccine than the disease. As a result unfounded stories of the dangers of vaccines easily tip public opinion against vaccination, with very serious consequences for child health.

Serious adverse reactions which are caused by vaccines, such as severe allergic reactions (anaphylaxis), are very rare. Most other vaccine reactions are minor and recover without treatment: fever, or pain, swelling and redness at the site of vaccination.

Monoclonal antibodies and blood clotting

PRODUCTION OF MONOCLONAL ANTIBODIES

Large quantities of a single type of antibody can be made using an ingenious technique.

- Antigens that correspond to a desired antibody are injected into an animal.
- B-cells producing the desired antibody are extracted from the animal.
- Tumour cells are obtained. These cells grow and divide endlessly.
- The B-cells are fused with the tumour cells, producing hybridoma cells that divide endlessly and produce the desired antibody.
- The hybridoma cells are cultured and the antibodies that they produce are extracted and purified.

The figure (right) shows a factory used for the industrial production of monoclonal antibodies. There are many ways in which monoclonal antibodies can be used. Two examples are described here.

Treatment of anthrax

Anthrax is a disease caused by a bacterium that produces toxins. It is often lethal, even when antibiotic treatments are given. Anthrax spores have sometimes been used deliberately to infect people and cause death.
Monoclonal antibodies are being developed which neutralize one of the toxins and therefore sustain the patient's life until their immune system produces antibodies naturally.

Diagnosis of malaria

Tests using monoclonal antibodies have been developed for many diseases, including malaria. Monoclonal antibodies are produced that bind to antigens in malarial parasites. A test plate is coated with the antibodies. A sample is left in the plate long enough for malaria antigens in the sample to bind to the antibodies. The sample is then rinsed off the plate. Any bound antigens are detected using more monoclonal antibodies with enzymes attached that cause a colour change. This is called an ELISA test. It can be used to measure the level of infection and to distinguish between different strains of malaria, either in humans or in mosquitoes.

BLOOD CLOTTING

When human tissue is injured and blood escapes from blood vessels, a semi-solid is formed from liquid blood to seal up the wound and prevent entry of pathogens. The semi-solid is called a **blood clot** and the process is called **clotting**.

Platelets have an important role in clotting. Platelets are small cell fragments that circulate with erythrocytes and leukocytes in the blood plasma. The clotting process begins with the release of clotting factors either from damaged tissue cells or from platelets. These clotting factors set off a series of reactions in which the product of each reaction is the catalyst of the next reaction. This system helps to ensure that clotting only happens when it is needed and it also makes it a very rapid process. In the last reaction fibrinogen, a soluble plasma protein is altered by the removal of sections of peptide that have many negative charges. This allows the remaining polypeptide to bind to others, forming long protein fibres called fibrin. Fibrin forms a mesh of fibres across wounds. Blood cells are caught in the mesh and soon form a semi-solid clot. If exposed to air the clot dries to form a protective scab, which remains until the wound has healed.

Final reactions in blood clotting

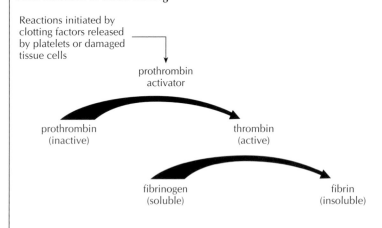

Fibrin and blood cells in a blood clot

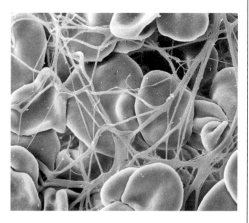

Muscles and joints

MOVEMENT IN HUMANS

The muscular-skeletal system and nervous system are responsible for movement in humans.

- **Muscles** provide the force needed for muscle contraction. They do this when they contract.
- **Tendons** attach muscles to bone.
- **Bones** provide a firm anchorage for muscles. They also act as levers, changing the size or direction of forces generated by muscles.
- **Ligaments** connect bone to bone, restricting movement at joints and helping to prevent dislocation.
- **Nerves** stimulate muscles to contract at a precise time and extent, so that movement is co-ordinated.

JOINTS

Junctions between bones are called joints.

- **Cartilage** reduces friction between bones where they meet.
- **Synovial fluid** lubricates the joint to reduce friction.
- **Joint capsule** seals the joint and holds in the synovial fluid.

The knee is a hinge joint. It allows considerable movement in one plane: bending (flexion) or straightening (extension), but little movement in the other two planes. In contrast, the hip joint allows movement in three planes (protraction /retraction, abduction/adduction and rotation). The figure (below) shows the structure of the elbow joint.

THE ELBOW JOINT

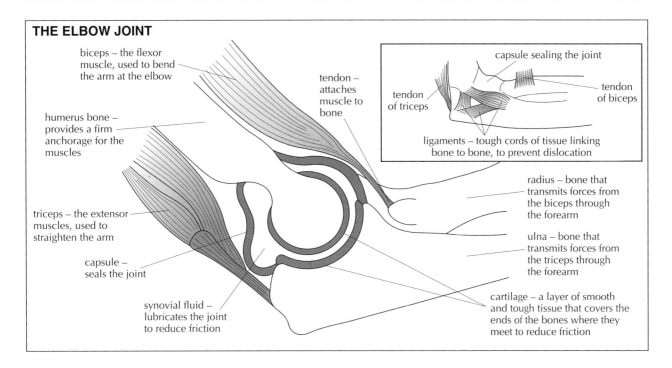

biceps – the flexor muscle, used to bend the arm at the elbow

humerus bone – provides a firm anchorage for the muscles

triceps – the extensor muscles, used to straighten the arm

capsule – seals the joint

synovial fluid – lubricates the joint to reduce friction

tendon – attaches muscle to bone

capsule sealing the joint

tendon of triceps

tendon of biceps

ligaments – tough cords of tissue linking bone to bone, to prevent dislocation

radius – bone that transmits forces from the biceps through the forearm

ulna – bone that transmits forces from the triceps through the forearm

cartilage – a layer of smooth and tough tissue that covers the ends of the bones where they meet to reduce friction

STRUCTURE OF STRIATED MUSCLE FIBRES

When viewed with a light microscope skeletal muscle is seen to consist of large multinucleate cells called **muscle fibres**. Within each muscle fibre are cylindrical structures called **myofibrils**. The myofibrils consist of repeating units called **sarcomeres**, which have light and dark bands. The light and dark bands extend across all the myofibrils in a muscle fibre, giving it a striated (striped) appearance. Around each myofibril is a special type of endoplasmic reticulum, called sarcoplasmic reticulum, visible in the electron micrograph (right). There are also mitochondria between the myofibrils.

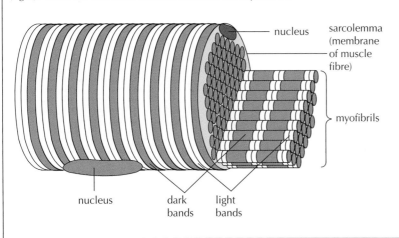

nucleus

sarcolemma (membrane of muscle fibre)

myofibrils

nucleus

dark bands

light bands

Muscle contraction

STRUCTURE OF A SARCOMERE

A sarcomere is a subunit of a myofibril. At either end is a Z line to which narrow actin filaments are attached. The actin filaments stretch inwards towards the centre of the sarcomere. Between them, there are thicker myosin filaments, which have heads that can bind to the actin. The part of the sarcomere containing myosin is the dark band and the part containing only actin filament is the light band. The figure (right) shows the structure of a sarcomere.

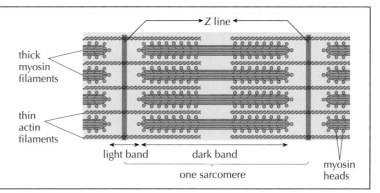

CONTRACTION OF SKELETAL MUSCLE

When a motor neurone stimulates a striated muscle fibre, calcium ions are released from the sarcoplasmic reticulum. The calcium causes binding sites on actin to be revealed, allowing myosin heads to bind. A cycle of events occurs, which causes actin filaments to slide inwards towards the centre of the sarcomeres. This makes the light bands narrower and the sarcomeres shorter – the muscle fibre contracts. The figure (below) shows the events that cause muscle contraction. The electron micrographs show striated muscle in relaxed (bottom left) and contracted (bottom right) states.

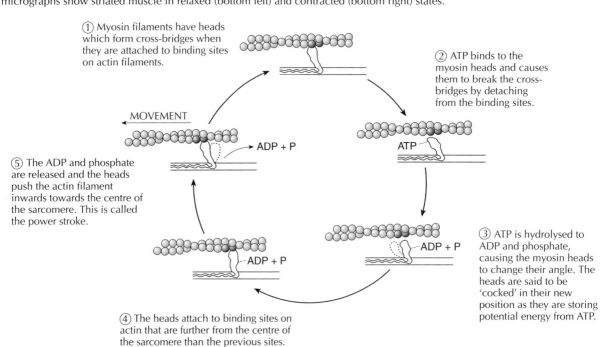

① Myosin filaments have heads which form cross-bridges when they are attached to binding sites on actin filaments.

② ATP binds to the myosin heads and causes them to break the cross-bridges by detaching from the binding sites.

③ ATP is hydrolysed to ADP and phosphate, causing the myosin heads to change their angle. The heads are said to be 'cocked' in their new position as they are storing potential energy from ATP.

④ The heads attach to binding sites on actin that are further from the centre of the sarcomere than the previous sites.

⑤ The ADP and phosphate are released and the heads push the actin filament inwards towards the centre of the sarcomere. This is called the power stroke.

MOVEMENT

ADP + P

ATP

ADP + P

ADP + P

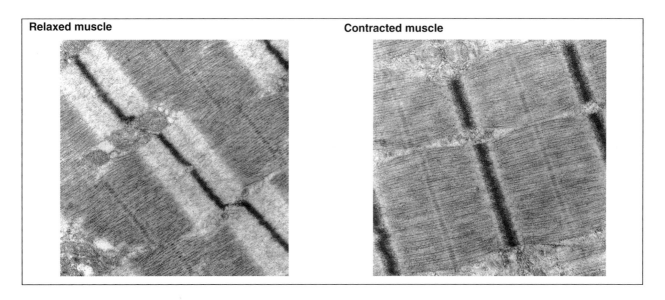

Relaxed muscle

Contracted muscle

Kidney structure and ultrafiltration

FUNCTIONS OF THE KIDNEY

The kidney has two functions – excretion and osmoregulation.

Excretion is the removal from the body of the waste products of metabolic pathways.

Osmoregulation is the control of the water balance of the blood, tissue or cytoplasm of a living organism.

THE STRUCTURE OF THE KIDNEY

The kidneys produce urine. The figure (below) shows the structure of the kidney. The cortex and medulla of the kidney contain many narrow tubes called nephrons. The figure (below) shows the structure of a nephron, together with the associated glomerulus.

Structure of a kidney in vertical section

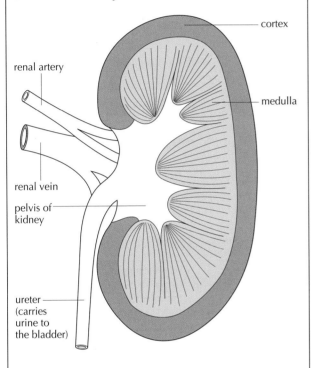

Structure of part of a glomerulus

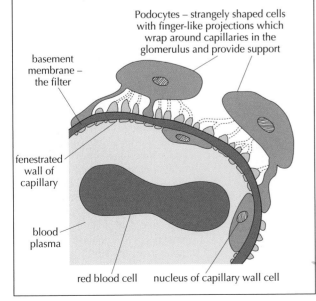

ULTRAFILTRATION IN THE GLOMERULUS

The function of the glomerulus is production of a filtrate from blood by a process called **ultrafiltration**. Part of the blood plasma escapes through the walls of all capillaries, but in the glomerulus 20% escapes, which is much greater than usual. There are two main reasons for this.

* The blood pressure is very high, because the vessel taking blood away from the glomerulus is narrower than the vessel bringing blood.
* The capillaries in the glomerulus are fenestrated – they have many pores through them.

These pores are large enough to allow any molecules through, but on the outside of the capillary wall is a basement membrane, composed of a gel of glycoproteins (below). The basement membrane acts as a filter as it only allows molecules with a molecular mass below 68 000 to pass through. It lets all substances in blood plasma through except plasma proteins. The fluid produced by ultrafiltration is collected by the Bowman's capsule and flows on into the proximal convoluted tubule.

Structure of the nephron

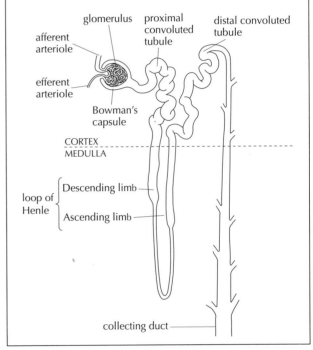

COMPARISON OF FLUIDS IN THE KIDNEY

The physiology of the kidney can be studied by comparing the content of blood flowing to and from the kidney with the content of glomerular filtrate and urine.

	Content (mg per 100ml)			
	Blood in renal artery	**Urine**	**Glomerular filtrate**	**Blood in renal vein**
Glucose	90	0	90	90
Urea	30	2000	30	24
Proteins	740	0	0	740

Glucose is often present in the urine of untreated diabetic patients. This is because the glucose concentration of blood rises much higher than 90 mg per 100 ml, so the pumps in the proximal convoluted tubule cannot reabsorb all the glucose that is filtered out in the glomerulus.

Urine production and osmoregulation

SELECTIVE RE-ABSORPTION IN THE PROXIMAL CONVOLUTED TUBULE

Large volumes of glomerular filtrate are produced – about 1 litre every 10 minutes by the two kidneys. As well as waste products, the filtrate contains substances that the body needs, which must be re-absorbed into the blood. Most of this **selective re-absorption** happens in the proximal convoluted tubule. The wall of the nephron consists of a single layer of cells. In the proximal convoluted tubule the cells have microvilli projecting into the lumen (right), giving a large surface area for absorption. Pumps in the membrane re-absorb useful substances by active transport, using ATP produced by mitochondria in the cells. All of the glucose in the filtrate is re-absorbed. About 80% of the mineral ions, including sodium is re-absorbed. Active transport of solutes makes the total solute concentration higher in the cells of the wall than in the filtrate in the tubule. Water therefore moves from the filtrate to the cells and on into the adjacent blood capillary by osmosis. About 80% of the water in the filtrate is re-absorbed, leaving 20% of the original volume to flow on into the loop of Henle.

Structure of the proximal convoluted tubule

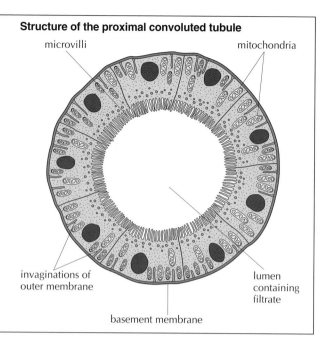

microvilli mitochondria

invaginations of outer membrane

lumen containing filtrate

basement membrane

THE ROLE OF THE LOOP OF HENLE

Glomerular filtrate flows deep into the medulla in descending limbs of the loops of Henle and then back out to the cortex in ascending limbs. Descending limbs and ascending limbs are opposite in terms of permeability. Descending limbs are permeable to water but not to sodium ions. Ascending limbs are permeable to sodium ions but not to water (right). Ascending limbs pump sodium ions from the filtrate into the medulla by active transport, creating a high solute concentration in the medulla. As the filtrate flows down the descending limb into this region of high solute concentration, some water is drawn out by osmosis. This dilutes the fluids in the medulla slightly. However the filtrate that leaves the loop of Henle is more dilute than the fluid entering it, showing that the overall effect of the loop of Henle is to increase the solute concentration of the medulla. This is the role of the loop of Henle – to create an area of high solute concentration in the cells and tissue fluid of the medulla. After the loop of Henle, the filtrate passes through the distal convoluted tubule, where the ions can be exchanged between the filtrate and the blood to adjust blood levels. It then passes into the collecting duct.

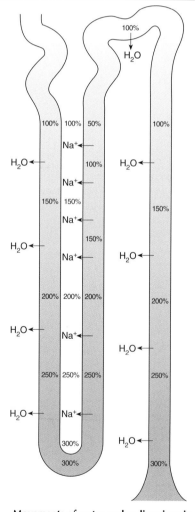

Movements of water and sodium ions in the loop of Henle and the collecting duct. Solute concentrations inside and outside the nephron are shown as a percentage of normal blood solute concentration

OSMOREGULATION IN THE COLLECTING DUCT

Osmoregulation is the control of water and solute levels. The collecting duct has an important role in osmoregulation. If the water content of the blood is too low, the pituitary gland secretes ADH. This hormone makes the cells of the collecting duct produce membrane channels called aquaporins, which makes the collecting duct permeable to water. As the filtrate passes down the collecting duct through the medulla, the high solute concentration of the medulla causes most of the water in the filtrate to be re-absorbed by osmosis. A small volume of concentrated urine is produced.

If the water content of the blood is too high, ADH is not secreted, aquaporins are broken down and the collecting duct becomes much less permeable to water. Little water is reabsorbed as the filtrate passes down the collecting duct and a large volume of dilute urine is produced. In this way the water content of the blood is kept within narrow limits. The urine produced by the collecting ducts drains into the renal pelvis and down the ureter to the bladder.

Spermatogenesis

Spermatogenesis is the production of spermatozoa.
Spermatozoa are usually simply called sperm.
Spermatogenesis occurs in the testes, in narrow tubes called
seminiferous tubules.
The figures (below and right) show the structure of testis
tissue, including the seminiferous tubules. The figure (bottom)
shows the processes involved in spermatogenesis.

Structure of testis tissue

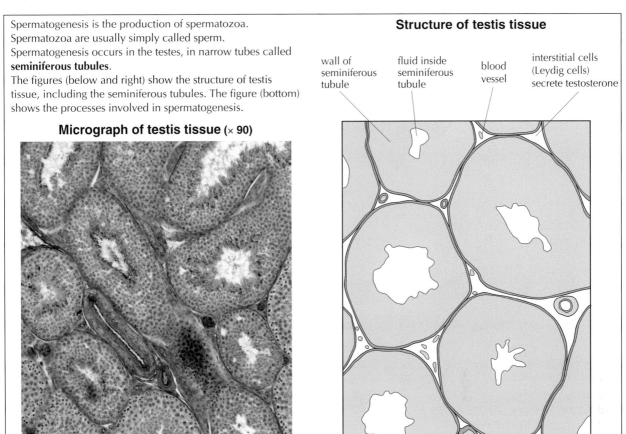

wall of seminiferous tubule

fluid inside seminiferous tubule

blood vessel

interstitial cells (Leydig cells) secrete testosterone

Micrograph of testis tissue (× 90)

STAGES OF SPERMATOGENESIS

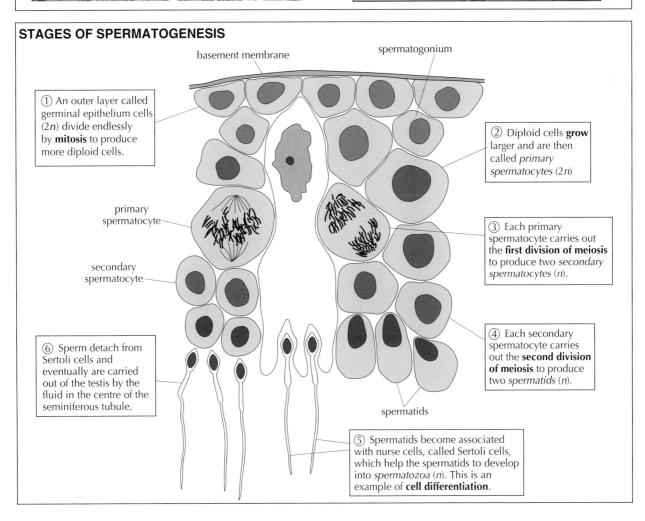

basement membrane

spermatogonium

① An outer layer called germinal epithelium cells (2*n*) divide endlessly by **mitosis** to produce more diploid cells.

② Diploid cells **grow** larger and are then called *primary spermatocytes* (2*n*)

primary spermatocyte

③ Each primary spermatocyte carries out the **first division of meiosis** to produce two *secondary spermatocytes* (*n*).

secondary spermatocyte

④ Each secondary spermatocyte carries out the **second division of meiosis** to produce two *spermatids* (*n*).

⑥ Sperm detach from Sertoli cells and eventually are carried out of the testis by the fluid in the centre of the seminiferous tubule.

spermatids

⑤ Spermatids become associated with nurse cells, called Sertoli cells, which help the spermatids to develop into *spermatozoa* (*n*). This is an example of **cell differentiation**.

Oogenesis

Oogenesis is the production of an ovum. Ova are often simply called eggs. Oogenesis occurs in the ovaries.
The figures below show the structure of ovary tissue.
The figure (bottom) shows the processes involved in oogenesis.

Micrograph of the ovary of a rabbit

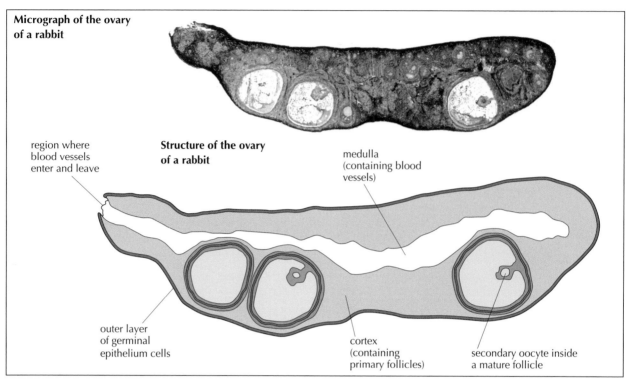

region where blood vessels enter and leave

Structure of the ovary of a rabbit

medulla (containing blood vessels)

outer layer of germinal epithelium cells

cortex (containing primary follicles)

secondary oocyte inside a mature follicle

STAGES OF OOGENESIS

③ Primary oocytes start the **first division of meiosis** but stop during prophase I. The primary oocyte and a single layer of follicle cells around form a primary follicle.

④ When a baby girl is born the ovaries contain about 400 000 primary follicles.

⑤ Every menstrual cycle a few primary follicles start to develop. The primary oocyte completes the first division of meiosis, forming two haploid nuclei. The cytoplasm of the primary oocyte is **divided unequally** forming a large secondary oocyte (n) and a small polar cell (n).

② Diploid cells **grow** into larger cells called primary oocytes ($2n$).

① In the ovaries of a female fetus, germinal epithelium cells ($2n$) divide by **mitosis** to form more diploid cells ($2n$).

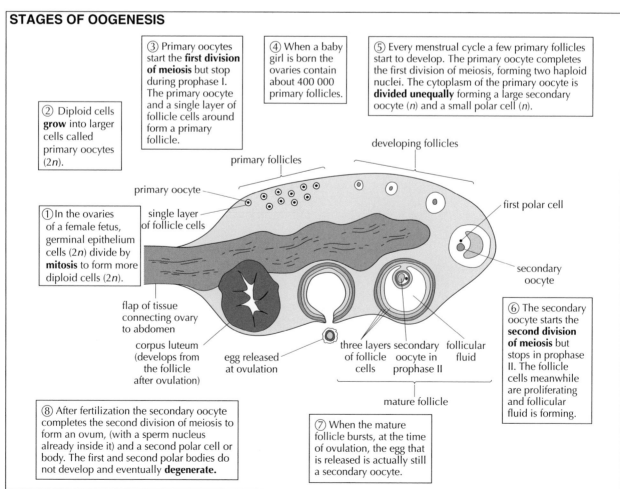

developing follicles

primary follicles

primary oocyte

single layer of follicle cells

first polar cell

secondary oocyte

flap of tissue connecting ovary to abdomen

corpus luteum (develops from the follicle after ovulation)

egg released at ovulation

three layers of follicle cells

secondary oocyte in prophase II

follicular fluid

mature follicle

⑥ The secondary oocyte starts the **second division of meiosis** but stops in prophase II. The follicle cells meanwhile are proliferating and follicular fluid is forming.

⑧ After fertilization the secondary oocyte completes the second division of meiosis to form an ovum, (with a sperm nucleus already inside it) and a second polar cell or body. The first and second polar bodies do not develop and eventually **degenerate.**

⑦ When the mature follicle bursts, at the time of ovulation, the egg that is released is actually still a secondary oocyte.

Gametes

STRUCTURE OF HUMAN SPERM

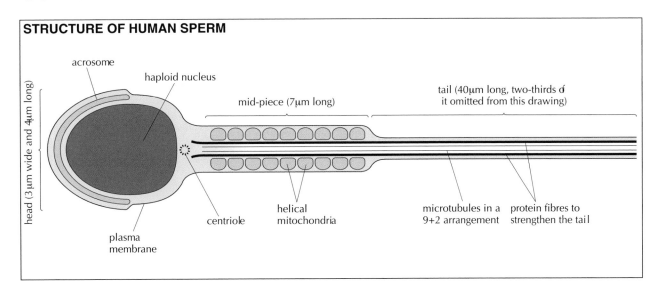

head (3 μm wide and 4μm long)

acrosome

haploid nucleus

mid-piece (7μm long)

tail (40μm long, two-thirds of it omitted from this drawing)

plasma membrane

centriole

helical mitochondria

microtubules in a 9+2 arrangement

protein fibres to strengthen the tail

HORMONAL CONTROL OF SPERMATOGENESIS

Three hormones are involved in the production of sperm.

Hormone	Source	Role
FSH	Pituitary gland	Stimulates primary spermatocytes to undergo the first division of meiosis, to form secondary spermatocytes
Testosterone	Interstitial cells in the testis	Stimulates the development of secondary spermatocytes into mature sperm
LH	Pituitary gland	Stimulates the secretion of testosterone by the testis

PRODUCTION OF SEMEN

Three structures help to produce semen – the epididymis, seminal vesicles and prostate gland

When sperm from the testis arrive in the epididymis, they are unable to swim. The sperm undergo a maturing process while they are stored in the epididymis and become able to swim. The two seminal vesicles and prostate gland produce and store fluids and expel them during ejaculation. The fluid mixes with the sperm and increases the volume of the ejaculate. The fluid from the seminal vesicles contains nutrients for the sperm including fructose. It also contains mucus which protects the sperm in the vagina. The fluid from the prostate gland contains mineral ions and is alkaline so protects the sperm from the acid conditions in the vagina.

STRUCTURE OF A HUMAN EGG

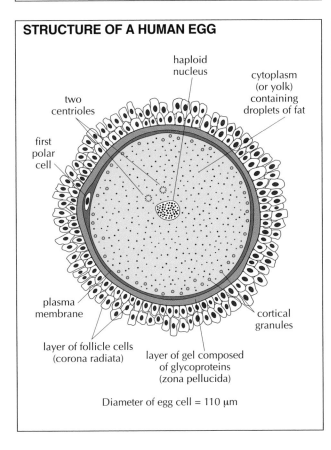

haploid nucleus

cytoplasm (or yolk) containing droplets of fat

two centrioles

first polar cell

plasma membrane

cortical granules

layer of follicle cells (corona radiata)

layer of gel composed of glycoproteins (zona pellucida)

Diameter of egg cell = 110 μm

COMPARING SPERMATOGENESIS WITH OOGENESIS

There are many similarities between the formation of sperm and eggs.
- Both start with proliferation of cells by mitosis.
- Both involve the cell growth before meiosis.
- Both involve the two divisions of meiosis.

The table below shows some of the differences.

Spermatogenesis	Oogenesis
Millions produced daily	One produced every 28 days
Released during ejaculation	Released on about day 14 of menstrual cycle by ovulation
Sperm formation starts during puberty in boys	The early stages of egg production happen during fetal development in females
Sperm production continues throughout the adult life of men	Egg production becomes irregular and then stops at the menopause in women
Four sperm are produced per meiosis	Only one egg is produced per meiosis

Fertilization

Summary of spermatogenesis

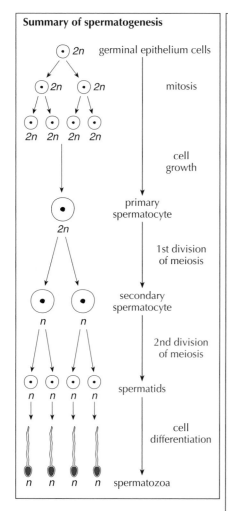

germinal epithelium cells

mitosis

cell growth

primary spermatocyte

1st division of meiosis

secondary spermatocyte

2nd division of meiosis

spermatids

cell differentiation

spermatozoa

Summary of oogenesis

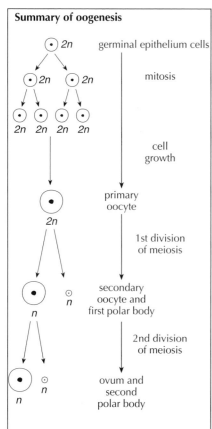

germinal epithelium cells

mitosis

cell growth

primary oocyte

1st division of meiosis

secondary oocyte and first polar body

2nd division of meiosis

ovum and second polar body

Stages in the fertilization of a human egg

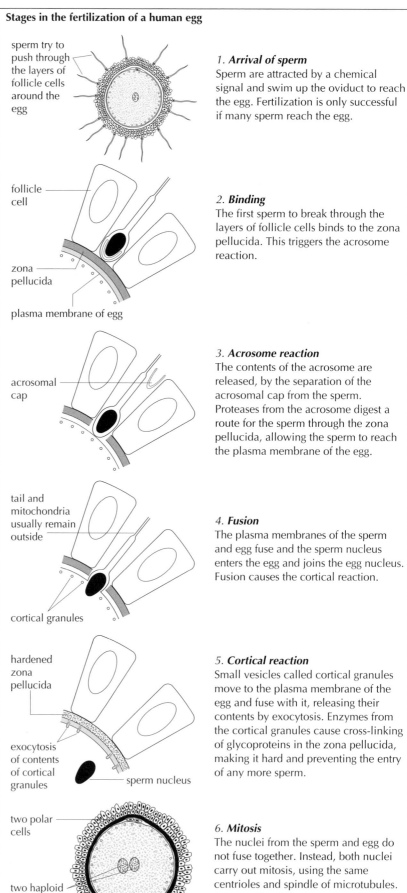

sperm try to push through the layers of follicle cells around the egg

follicle cell

zona pellucida

plasma membrane of egg

acrosomal cap

tail and mitochondria usually remain outside

cortical granules

hardened zona pellucida

exocytosis of contents of cortical granules

sperm nucleus

two polar cells

two haploid nuclei from the sperm and the egg

1. *Arrival of sperm*
Sperm are attracted by a chemical signal and swim up the oviduct to reach the egg. Fertilization is only successful if many sperm reach the egg.

2. *Binding*
The first sperm to break through the layers of follicle cells binds to the zona pellucida. This triggers the acrosome reaction.

3. *Acrosome reaction*
The contents of the acrosome are released, by the separation of the acrosomal cap from the sperm. Proteases from the acrosome digest a route for the sperm through the zona pellucida, allowing the sperm to reach the plasma membrane of the egg.

4. *Fusion*
The plasma membranes of the sperm and egg fuse and the sperm nucleus enters the egg and joins the egg nucleus. Fusion causes the cortical reaction.

5. *Cortical reaction*
Small vesicles called cortical granules move to the plasma membrane of the egg and fuse with it, releasing their contents by exocytosis. Enzymes from the cortical granules cause cross-linking of glycoproteins in the zona pellucida, making it hard and preventing the entry of any more sperm.

6. *Mitosis*
The nuclei from the sperm and egg do not fuse together. Instead, both nuclei carry out mitosis, using the same centrioles and spindle of microtubules. A two-cell embryo is produced.

Pregnancy and childbirth

FERTILIZATION AND EARLY EMBRYO DEVELOPMENT

If a couple want to have a child, they have sexual intercourse without using any method of contraception. The biological term for sexual intercourse is **copulation**. During copulation, semen is ejaculated into the vagina. Sperm swim through the cervix, up the uterus and into the oviducts. If there is an egg in the oviducts, a sperm can fuse with it to produce a zygote. The fusion of an egg with a sperm is called **fertilization**.

The zygote produced by fertilization in the oviduct is a new human individual. It starts to divide by mitosis to form a 2-cell embryo, then a 4-cell embryo (right) and so on until a hollow ball of cells called a blastocyst is formed. While these early stages in the development of the embryo are happening, the embryo is transported down the oviduct to the uterus. When it is about 7 days old, the embryo implants itself into the wall of the uterus, where it continues to grow and develop.

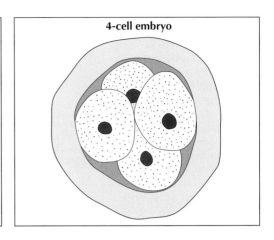

4-cell embryo

DEVELOPMENT OF THE FETUS

By the time that embryo is about 8 weeks old, it starts to develop bone tissue and is known from then onwards as a **fetus**. The fetus develops a placenta and an umbilical cord (left). The placenta is a disc-shaped structure, with many projections called placental villi embedded in the uterus wall. In the placenta the blood of the fetus flows close to the blood of the mother in the uterus wall. Materials are exchanges between maternal and fetal blood. For example, oxygen passes from maternal to fetal blood and carbon dioxide passes from fetal to maternal blood. The fetus also develops around itself an amniotic sac containing amniotic fluid. The fetus floats in this amniotic fluid and is supported by it. The delicate tissues of the fetus are protected from injury by the amniotic fluid, which acts as a shock absorber. This is needed if an everyday event or an accident causes an impact to the mother's abdomen.

A sample of fluid can be taken from the amniotic sac by inserting a hypodermic needle through the abdomen wall. This procedure is known as **amniocentesis**. The fluid contains fetal cells which can be cultured to make them divide. The chromosomes of the dividing cells can be examined to test for chromosomal abnormalities such as Down's syndrome.

Female reproductive system during pregnancy

- amniotic sac
- developing fetus
- uterus wall
- vagina
- amniotic fluid
- placenta
- umbilical cord
- cervix

CHILDBIRTH

Through the 9 months of pregnancy, the hormone progesterone ensures that the uterus develops and sustains the growing fetus. The level of progesterone in the mother becomes increasingly high. The end of pregnancy is signalled by a fall in progesterone level. This allows the mother's body to secrete another hormone – oxytocin. Oxytocin causes the muscle in the uterus wall to contract. Uterine contractions stimulate the secretion of more oxytocin. The uterine contractions therefore become stronger and stronger. This is an example of **positive feedback**.

While the muscle in the wall of the uterus is contracting, the cervix relaxes and becomes wider. The amniotic sac bursts and the amniotic fluid is released. Finally, often after many hours of contractions, the baby is pushed out through the cervix and the vagina. The umbilical cord is cut and the baby begins its independent life. Contractions continue for a time until the placenta is expelled as the afterbirth.

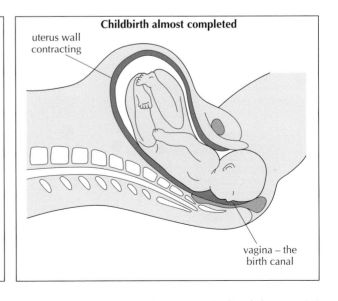

Childbirth almost completed

- uterus wall contracting
- vagina – the birth canal

Structure and function of the placenta

HORMONAL CONTROL OF PREGNANCY

Estrogen and progesterone are needed throughout pregnancy to stimulate the development of the uterus lining. During the first few days after ovulation the corpus luteum secretes these hormones whether or not there has been fertilization. After implanting in the uterus wall, the embryo starts to secrete a hormone called HCG (human chorionic gonadotrophin). HCG prevents degeneration of the corpus luteum, which would happen at the end of a menstrual cycle. HCG stimulates the corpus luteum to grow and to continue secretion of estrogen and progesterone. This is essential to allow the pregnancy to continue. By the middle of the pregnancy, the corpus luteum starts to degenerate, but by then cells in the placenta are secreting estrogen and progesterone and these cells secrete increasing amounts until the end of the pregnancy.

STRUCTURE AND FUNCTION OF THE PLACENTA

The figure (below) shows the structure and functions of the placenta.
The figure (bottom) shows how materials are exchanged between maternal and fetal blood at the surface of villi in the placenta.

Structure of the placenta

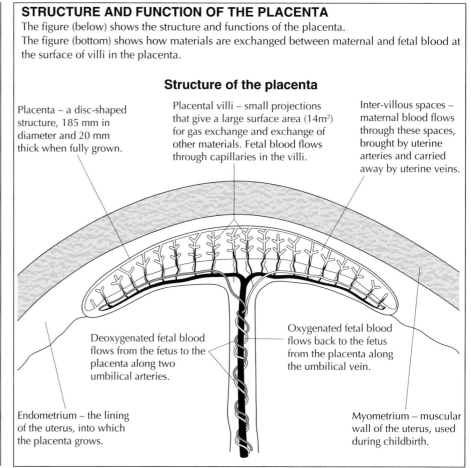

Placenta – a disc-shaped structure, 185 mm in diameter and 20 mm thick when fully grown.

Placental villi – small projections that give a large surface area (14m^2) for gas exchange and exchange of other materials. Fetal blood flows through capillaries in the villi.

Inter-villous spaces – maternal blood flows through these spaces, brought by uterine arteries and carried away by uterine veins.

Deoxygenated fetal blood flows from the fetus to the placenta along two umbilical arteries.

Oxygenated fetal blood flows back to the fetus from the placenta along the umbilical vein.

Endometrium – the lining of the uterus, into which the placenta grows.

Myometrium – muscular wall of the uterus, used during childbirth.

EXCHANGE OF MATERIALS ACROSS THE PLACENTA

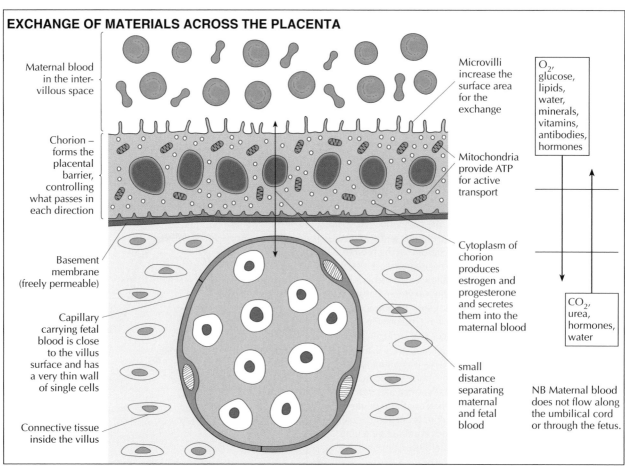

Maternal blood in the inter-villous space

Microvilli increase the surface area for the exchange

O$_2$, glucose, lipids, water, minerals, vitamins, antibodies, hormones

Chorion – forms the placental barrier, controlling what passes in each direction

Mitochondria provide ATP for active transport

Basement membrane (freely permeable)

Cytoplasm of chorion produces estrogen and progesterone and secretes them into the maternal blood

Capillary carrying fetal blood is close to the villus surface and has a very thin wall of single cells

CO$_2$, urea, hormones, water

small distance separating maternal and fetal blood

NB Maternal blood does not flow along the umbilical cord or through the fetus.

Connective tissue inside the villus

EXAM QUESTIONS ON TOPIC 11

1 a) Define *excretion*.   [2]

   b) Compare the composition of blood plasma and urine, by giving two differences
       in the table below.   [2]

Blood in the renal artery	Blood in the renal vein

   c) Explain briefly the function of the loop of Henle in the human kidney.   [2]

   d) Deduce which part of the kidney has been damaged if protein is found in the urine.   [1]

2 The electron micrograph below shows part of a myofibril, taken from a skeletal muscle. The parts marked
   M contain myosin filaments. Three other regions are labelled **I**, **II** and **III**.

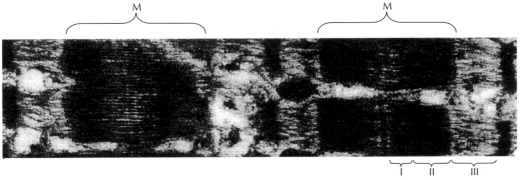

[Source: Dr G. Newman, EM Unit, University of Wales College of Medicine]

   a) (i)   State one type of filament, apart from myosin, which is present in myofibrils.   [1]

       (ii) Identify in which of the regions labelled I, II and III these other filaments can be found.   [1]

   b) The myofibril is partly contracted. Deduce which of the regions would increase in length if

       (i)   the myofibril contracted more   [1]

       (ii) the myofibril relaxed   [1]

3 a) Compare the structure of human sperm and eggs.   [4]

   b) Compare the role of FSH in men and women.   [3]

   c) Compare the roles of LH and HCG in women.   [3]

Components of the human diet

NUTRIENTS IN THE HUMAN DIET

Nutrients are chemical substances found in foods that are used in the human body.

Many nutrients are needed in the human diet:
amino acids, **fatty acids**, **minerals**, **vitamins** and **water**.

Carbohydrates are almost always present in human diets but specific carbohydrates are not essential.

Minerals and vitamins are both needed in small quantities, but they are chemically very different:

- minerals are chemical elements that are obtained in an ionic form, for example sodium as Na^+ ions and phosphorus as PO_4^{3-} ions
- vitamins are organic compounds that cannot be synthesized by the body.

RECOMMENDED VITAMIN C INTAKES

The recommended daily intake of a mineral or vitamin is the minimum amount that should be consumed per day to ensure health. Two types of experiment have been done to determine this amount for vitamin C.

1. A small mammal called a guinea pig has been used, with groups of guinea pigs fed diets with different levels of vitamin C for a trial period. The level of vitamin C in the blood plasma and urine of the experimental animals was then measured. Also the strength of collagen in bone and skin was measured, to test for signs of scurvy (vitamin C deficiency). The results can be used to estimate the amount of vitamin C that humans require per kilogram of body tissue.

2. Experiments were done in Britain during the Second World War, using volunteers who were conscientious objectors to military service. For six weeks all the volunteers were given a diet lacking vitamin C plus a daily supplement of 70mg. The volunteers were then divided into three groups and given 70mg, 10mg or no vitamin C per day, for eight months. They were given skin cuts, to test for wound healing and were checked for other signs of scurvy. The 10mg and 70mg groups did not develop scurvy, but all those in the 0mg group did.

The results of the human trial suggest that 10mg of vitamin C per day is sufficient, but in most countries much higher recommended daily intakes have been set. There are two main reasons for this.

- To give a safety margin so that the risk of scurvy is minimized
- To allow for variations between individuals in their general health and in their ability to absorb and use vitamin C

Some scientists have suggested that the recommended level should be increased, because vitamin C may give protection against upper respiratory tract infections (colds). For example, Linus Pauling, a Nobel prize-winning American chemist, advocated levels of 1000mg or more of vitamin C per day. There is little scientific evidence to back up the theory that this gives protection against colds. There is evidence that if the body adjusts to high rates of intake by excreting the excess and intake drops back to normal levels, this excretion continues, causing scurvy to develop – this is called rebound malnutrition.

Daily intake of about 50mg is probably sufficient.

VITAMIN D AND THE BALANCING OF RISK

If there is insufficient vitamin D in the body, calcium is not absorbed from food in the gut in large enough quantities. Symptoms of calcium deficiency can develop (rickets).

Vitamin D is contained in oily fish, eggs, milk, butter, cheese and liver. Plant products do not contain vitamin D, but it is usually added during the manufacture of soya milk, margarine and breakfast cereals.

Unusually for a vitamin, it can be synthesised in the skin, but this only happens with ultra-violet light (UV). The intensity of UV is too low in winter in high latitudes for much vitamin D to be synthesized, but the liver can store enough during the summer to avoid a deficiency in winter.

Even when there is bright sunlight, three things can prevent absorption of enough ultra-violet light by the skin for adequate synthesis of vitamin D:

- avoiding exposure to sunlight e.g. staying indoors
- covering most of the skin with clothing
- applying sun creams to block ultra violet light

These things are all ways of reducing the risk of malignant melanoma – a form of skin cancer that can be fatal. This is because UV can cause the mutations that turn skin cells into tumour cells. A delicate balance must therefore be struck, between over-exposure to UV and an excessive risk of malignant melanoma, and under-exposure, which brings the risk of vitamin D deficiency.

IODINE AND DIETARY SUPPLEMENTATION

Artificial nutrient supplementation is used when a diet contains insufficient quantities of a nutrient. The nutrient can be added to a food, or can be supplied in a pure form. Iodine is one of the best examples of the benefits of artificial dietary supplementation.

Iodine is needed for the synthesis of the hormone thyroxin, by the thyroid gland. An obvious symptom of iodine deficiency disorder (IDD) is swelling of the thyroid gland in the neck, called goitre (see figure). IDD also has some less obvious but very serious consequences. If women are affected during pregnancy, their children are born with permanent brain damage. 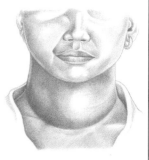 If children suffer from IDD after birth, their mental development and intelligence are impaired.

In 1998 UNICEF estimated that 43 million people worldwide had brain damage due to IDD and 11 million of these had a severe condition called cretinism. 40% of the world's population was estimated to show some mental impairment because of IDD, with highest rates in areas where the soil used to grow crops and the drinking water contain little iodine.

At the World Children's Summit in 1990 a campaign was started to eliminate IDD by adding it in small quantities to salt sold for human consumption. This is a highly effective way of preventing IDD at a cost of only about 5 cents per person per year. By the year 2000 iodized salt was reaching more than 3.3 billion people. If the campaign to provide iodized salt throughout the world is successful, no children in the future will suffer from IDD.

Amino acids and fatty acids

AMINO ACIDS IN THE HUMAN DIET

Twenty different amino acids are needed to make proteins. Humans can make about half of these in sufficient quantities by conversion from other nutrients in the diet. These are the **non-essential amino acids**. A shortage of any of the other amino acids in the diet causes **protein deficiency malnutrition**. Often this deficiency disease is caused by a general shortage of protein in the diet, which results in a lack of most or all of the essential amino acids.

One of the most obvious consequences of protein deficiency malnutrition is the swelling of the abdomen, often seen in photographs of children in areas of famine. The swelling is caused by tissue fluid retention, or oedema. In a well-nourished person, plasma proteins in the blood cause all tissue fluid to be reabsorbed in blood capillaries. If these proteins cannot be synthesized, because of a lack of one or more of the essential amino acids, then fluid builds up in the tissues, including the abdomen.

Protein deficiency malnutrition has many other effects, as proteins have so many roles in the body. The overall effect is that growth is retarded and if energy is also lacking in the diet there can be wasting – loss of body mass.

PHENYLKETONURIA

Phenylalanine is an essential amino acid, but tyrosine is non-essential because it can be synthesised from phenylalanine. This reaction is catalysed by the enzyme tyrosine hydroxylase. If this enzyme is cannot be produced, phenylalanine cannot be converted to tyrosine, so levels of phenylalanine in the blood rise, with harmful consequences. Mental and physical development is retarded in children, if the disease is not treated.

$$phenylalanine \xrightarrow{\text{tyrosine hydroxylase}} tyrosine$$

High levels of phenylalanine in the blood are a symptom of the genetic disease **phenylketonuria**. In this disease, there is a deficiency or complete lack of the enzyme tyrosine hydroxylase, because of a mutation of the gene coding for the enzyme. In a fetus, the levels of phenylalanine are kept down by the mother's metabolism, but after birth the levels rise and soon have harmful effects. In many countries, phenylalanine levels are tested by a simple blood test soon after birth, allowing very early diagnosis of phenylketonuria. This allows a special diet to be fed, containing low levels of phenylalanine, preventing most if not all of the harmful effects of the disease.

STRUCTURE OF FATTY ACIDS

All fatty acids have a carboxyl group (COOH) and a hydrocarbon chain. Fatty acids are variable in the bonding between the carbon atoms and the number of hydrogen atoms bonded to the carbons.

Saturated – all of the carbon atoms in the chain are connected by single covalent bonds so the number of hydrogen atoms bonded to the carbons cannot be increased.
Unsaturated – one or more double bonds between carbon atoms in the chain, so more hydrogen could be bonded to the carbons if a double bond was replaced by a single bond.

Unsaturated fatty acid
(naturally occurring ones have more carbon atoms)

Monounsaturated – only one double bond.
Polyunsaturated – two or more double bonds.
The position of the nearest double bond to the CH_3 terminal is significant. In omega-3 fatty acids, it is the third bond from CH_3 whereas in omega-6 fatty acids it is the sixth.
Cis unsaturated – hydrogen atoms are bonded to carbon atoms on the same side of a double bond.
Trans unsaturated – hydrogen atoms are bonded to carbon atoms on opposite sides of a double bond.

cis

trans

FATTY ACIDS AND HEALTH

Many claims have been made about the health consequences of the various types of fatty acid but the evidence for many of these claims is scanty.

Saturated fatty acids
- There is a positive correlation between diets with high levels of saturated fatty acids and CHD mortality.
- Correlation is not proof of cause – it could for example be the low fibre content of most diets high in saturated fat that causes CHD.

Cis-monounsaturated fatty acids
- Rates of CHD are typically low in people who eat a Mediterranean type diet, rich in olive oil, which contains cis-monounsaturated fatty acids.
- Other factors vary between people, apart from the amount of cis-monounsaturated fatty acids in their diets, including genetic factors, and these could explain differences in rates of CHD.

Omega-3 fatty acids
- Much of the tissue of the eye and brain is made up of long-chain fatty acids. These are synthesized in the body from essential fatty acids, including omega-3 fatty acids. It seems reasonable that a deficiency of these would impair brain and eye development.
- There is no clear evidence that omega-3 dietary supplements improve brain and eye development.

Trans fatty acids
- There is a positive correlation between diets with high levels of trans fatty acids and CHD. Analysis has shown that other factors cannot explain the correlation, leaving trans fatty acids as the only risk factor.
- In autopsies after deaths from CHD, most of the fat in arterial plaque has been found to be trans fat.
- Consumption of trans fats is associated with raised LDL levels and reduced HDL levels, which are associated with increased rates of CHD.

Energy in human diets

ENERGY CONTENTS
The three types of nutrient that supply most energy in human diets are carbohydrates, fats (lipids) and proteins. Of these, fats provide the most energy per gram. Carbohydrates and proteins have a similar energy content.

Nutrient	Energy content per 100g
Carbohydrate	1760 kJ
Fat	4000 kJ
Protein	1720 kJ

DIFFERENCES IN ENERGY SOURCES
Diets vary around the world, especially between ethnic groups that eat traditional diets. Usually a few foods are the main energy sources in the diet and these foods are eaten in large quantities.
- Rice – in tropical and temperate areas, for example in China and Japan
- Wheat – in areas with a temperate climate, for example in the Ukraine
- Cassava – in high rainfall areas in the tropics, for example the Yoruba tribe in Nigeria
- Fish – where crop growth is impossible, for example the Inuit tribe in the far north of America
- Meat – in ethnic groups with a nomadic lifestyle, for example the Maasai of Kenya.

ENERGY SOURCES AND HUMAN HEALTH
There are health consequences of diets rich in carbohydrates, fats and proteins.
Carbohydrates
Consumption of large amounts of sugar can increase the risk of obesity, Type II diabetes and tooth decay.
Consumption of large amounts of starch can cause obesity, but there is little evidence of other health problems, especially if the starch is in a formulation in which it is slowly digested, preventing rapid glucose absorption into the blood. Dietary fibre is mostly complex indigestible carbohydrate. The health benefits of eating it are described on page 114.
Fats
Consumption of fats in large quantities carries a significant risk of obesity. It has also been known for many years that there is positive correlation between fat intake and the risk of death from coronary heart disease (CHD). The type of fat is very significant, with trans fats associated with the greatest risk; saturated fats also probably increase the risk of CHD. Causes of CHD are complex and genetic factors and other dietary factors also have major influences.
Proteins
Large amounts of protein are sometimes consumed, especially as a part of some slimming diets that are intended to reduce body mass. Although controlled trials have not been done, there may be some associated health risks, especially with animal protein. These include kidney stones (composed of uric acid), gout, reduced kidney function in people who already have impaired kidney function, and loss of calcium in urine, increasing the risk of osteoporosis.

APPETITE CONTROL
The brain has an appetite control centre. It is located in the hypothalamus. Its role is to make us feel satiated when we have eaten enough food. The appetite control centre does this when it receives hormonal stimuli:
- insulin, secreted by the pancreas when blood glucose levels are high
- PYY_{3-36} secreted by the small intestine, when there is food in it
- leptin secreted by adipose tissue, with more secreted as amounts of stored fat increase.

The role of the appetite control centre is very important. People whose appetite control centre does not function properly find it much harder to avoid obesity.

BODY MASS INDEX
It is not possible to assess whether a person's body mass is at a healthy level simply by weighing them, because of natural variation in size between adults. Instead, body mass index is calculated. The units for BMI are kg/m^2.

$$BMI = \frac{\text{mass in kilograms}}{(\text{height in metres})^2}$$

The table below can be used to draw conclusions from a person's BMI.

Body mass index	Conclusion
below 18.5	underweight
18.5–24.9	normal weight
25.0–29.9	overweight
30.0 or more	obese

CLINICAL OBESITY
When a doctor diagnoses that a patient is obese, it is called clinical obesity. The World Health Organization has reported an obesity epidemic, with rates rising rapidly in some countries. Over 300 million adults worldwide are clinically obese. The reasons for this are complex and include these factors:
- foods with a high content of fat and/or sugar are cheap and widely available and smaller quantities of low energy and high fibre foods are eaten
- economic growth and cheaper foods have allowed larger portion sizes to be served
- more people are using automated means of transport, such as cars or buses, and fewer people are walking or using other active means of transport
- many people now have physically undemanding jobs, for example in offices, instead of labouring work, for example on farms
- many tasks that were done in the home by hand are now done by a machine
- the most popular pastimes have become less active, for example watching television or playing computer games, instead of active games or sports.

In summary, more food is being eaten, but less of the energy in it is being used up in daily activities.

Issues in nutrition (Part 1)

HUMAN MILK AND ARTIFICIAL MILK

Human milk is produced by a mother's breasts after birth and while the baby is still suckling. Artificial milk, or infant formula, is produced with a content as similar as possible to human milk, but it cannot be identical in composition. The differences are summarized below.

	Human milk	Artificial milk
Carbohydrate	lactose	lactose OR glucose polymers
Protein source	65% human whey proteins 35% casein	18% bovine whey and 82% bovine casein OR soya proteins
Fatty acids	human butterfat	palm, coconut, soy or safflower oils
Antibodies	antibodies present in the first milk – colostrum	no antibodies for fighting human diseases are present

BENEFITS OF BREAST-FEEDING

Most mothers are advised to breast-feed (nurse) their babies, rather than bottle-feed with artificial milk, because of the benefits:
- breast-feeding avoids the allergies to proteins in cows' milk or soya that can develop when babies receive artificial milk
- breast-feeding promotes bonding between mother and baby
- frequent breast-feeding acts as a natural birth-control method, reducing the chance of conception while the mother is lactating and therefore allowing more time between the birth of one child and the next
- breast milk is naturally sterile so is safer in areas where it is impossible to sterilize water used to prepare artificial milk
- milk production helps mothers to lose weight after pregnancy.

ANOREXIA NERVOSA

Anorexia nervosa is a disease that mostly affects girls and women. It usually starts in the mid-teens when the emotional and psychological changes of adolescence are occurring. It has complex causes, making the condition a challenge for all those involved. The consequences for friends and family include anxiety about the physical harm that the condition causes, feelings of guilt about difficult relationships, powerlessness when treatment seems to be failing, and hurt at the desire for isolation shown by many people with anorexia. Because people with anorexia do not eat enough carbohydrate or fat for use in cell respiration, protein is broken down. Muscles lose mass and become weaker, with feelings of fatigue. Hair becomes more brittle and thinner and there can be hair loss. The skin becomes dry and bruises easily, with a growth of fine hair all over the body. Blood pressure drops, with a slow heart rate and poor circulation. Menstrual cycles often stop, with no periods or ovulation, making girls with anorexia infertile.

TYPE II DIABETES

This is the type of diabetes that usually develops in adults rather than children and is not treated by injections of insulin.

1. **Causes**

Although beta cells in the pancreas still secrete insulin in response to high blood glucose levels, body cells become less responsive to the insulin. The causes of this are not entirely understood, but seem to be associated with increased blood concentrations of fatty acids. The following factors all increase the risk:
- diets rich in fat, and low in fibre
- obesity, due to overeating and lack of exercise
- genetic factors, which affect fat metabolism.

These risk factors vary between ethnic groups and there is therefore huge variation in rates of Type II diabetes, from less than 2% in China to 50% among the Pima Indians.

2. **Symptoms**

The symptoms of Type II diabetes are usually mild and sometimes develop very gradually over a period of years, so it is not always diagnosed quickly. These are the main symptoms that are used to diagnose the condition:
- elevated levels of blood glucose
- glucose in the urine – this can be detected by a simple test
- tiredness, increased appetite and loss of body mass
- needing to excrete urine frequently, due to production of large volumes of urine
- dehydration and thirst – from loss of water in urine.

3. **Dietary advice**

Changes to the diet are an obvious way of trying to control Type II diabetes. Advice usually includes:
- reducing the intake of saturated fats
- reducing the intake of sugar, especially in sweets (candy), snack foods and drinks
- eating more foods that are high in fibre, including vegetables and fruit
- eating regular small meals throughout the day, each meal including moderate amounts of carbohydrate, to prevent high blood sugar levels after a large meal
- eating carbohydrates with a low glycemic index (GI), because they are digested and absorbed slowly. The graph (below) shows the effects on blood glucose levels of eating high GI and low GI foods.

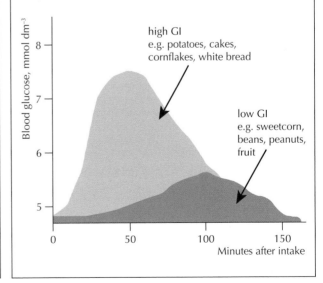

Issues in nutrition (Part 2)

ETHICS OF EATING ANIMAL PRODUCTS

Many people choose what to eat, based on likes and dislikes, availability and cost. Some people also have ethical reasons for not eating certain foods.

Meat

Animals have to be killed to obtain meat, usually after rearing them on a farm.

- Is it right for one animal to take the life of another animal to obtain food?
- Is the pain caused to animals during transport and slaughter justifiable?
- Is the suffering of animals reared for meat in unnatural and crowded conditions justifiable?

Milk

Cows and other mammals produce milk after giving birth. This milk can be used for human consumption if the calf or young mammal is separated from its mother soon after birth.

- Is the huge milk production of the cows that have been bred acceptable, given that it is often associated with health problems and a short life expectancy?
- Is the suffering of cows whose calves are taken away from them soon after birth justifiable?
- Is it acceptable to make cows have calves in order to stimulate milk production, when these calves will almost certainly have to be killed eventually?

Eggs

Most eggs come from hens (female birds) that have been specially bred for prolific egg production.

- Is it acceptable to breed and keep hens that produce far greater numbers of eggs than their wild relatives?
- Is the suffering of egg-laying hens kept in unnatural conditions justifiable – either in small cages or in artificially large groups in most free-range systems?
- Is it acceptable to kill male chicks at 1–3 days old because they do not lay eggs?

Honey

Bees are kept in hives and surplus honey is removed when available.

- Is it justifiable to take honey from bees that have stored it for their own use within the bee colony?
- Is it acceptable to keep bees in an area where the bees will compete with wild insects that forage on nectar from flowers?

CHOLESTEROL AND CHD

Cholesterol is a steroid and is mainly found in animal products. It is an essential component of membranes. Some investigations have shown that as the amount of cholesterol in the blood plasma rises, the risk of death from CHD (coronary heart disease) increases. A 10% increase in blood cholesterol is associated with a 30% increase in the risk of death from CHD. Other studies have suggested that total blood cholesterol is less significant than levels of cholesterol in LDL.

Nevertheless, it seems reasonable that reducing the amount of cholesterol in the diet should cut blood plasma concentrations, lowering the risk of CHD. In practice, dietary cholesterol only has a small effect on blood cholesterol levels so the effects of reducing dietary cholesterol are likely to be minimal.

Cholesterol can be synthesized by the liver and the rate of this varies as a result of genetic differences. In some families, high blood cholesterol levels are very common, even with diets very low in cholesterol.

The main correlation between diet and blood cholesterol levels is with saturated fat intake. As dietary saturated fat increases, both LDL and total blood cholesterol levels tend to increase. CHD rates are also correlated positively with saturated fat intake. It is not clear what, if any, the causal links are, but most physicians advise reducing saturated fat intake, to try to reduce blood cholesterol and the risk of CHD.

THE IMPORTANCE OF FIBRE

Fibre is material that cannot be digested in the small intestine. Cellulose from plant cell walls is the main component of dietary fibre, but there are others including chitin from fungi and crustaceans.

Many investigations have shown that fibre helps to prevent constipation, by increasing the bulk of material in the large intestine. There are other possible advantages, but the evidence for these is weaker.

- Fibre might help to prevent obesity by increasing the bulk in the stomach, which reduces the desire to eat more food.
- Fibre may reduce the risk of diseases of the large intestine including appendicitis, cancer and hemorrhoids.
- Fibre might slow the rate of sugar absorption and so help the prevention and treatment of diabetes.

FOOD MILES AND FOOD TRANSPORT

Food miles are simply a measure of how far a food item has been transported from where it was produced to where it is eaten. Much food is now transported hundreds of kilometres by road or rail, or thousands of kilometres by air. This causes air pollution, traffic congestion and the release of greenhouse gases.

At the other extreme, if urbanization was reversed and food was grown where we live, no energy would have to be used to transport the food.

Some consumers now refuse to buy foods with high food miles, hoping that supermarkets and shops will start selling locally produced food instead.

Other consumers are not concerned about food miles and instead want continuity of supply throughout the year and maximum choice of world foods.

Some environmentalists point out that there are other energy costs in food production, such as production and application of fertilizers. As an example, this might make that the overall energy costs of lamb produced using low energy input systems in New Zealand and transported around the world by sea, lower than the energy costs of locally produced lamb.

During famines, transport of food is justifiable on humanitarian grounds, whatever the food miles.

EXAM QUESTIONS ON OPTION A – HUMAN NUTRITION AND HEALTH

A1 The nomogram below shows the relationship between mass in kilograms, height in centimetres and body mass index for adults.

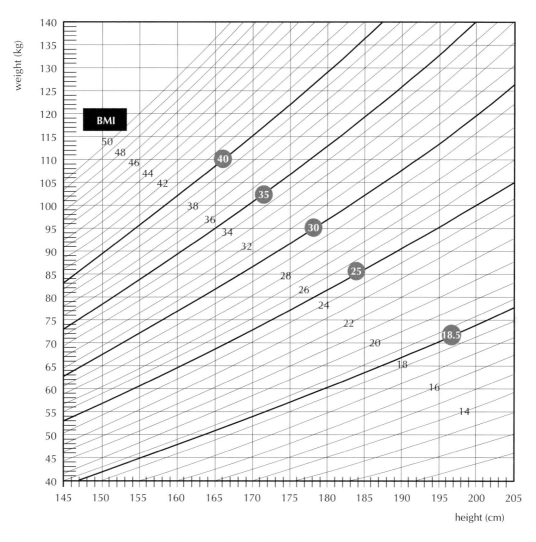

a) Use the nomogram to estimate the body mass index of adults with

   (i) a mass of 70 kg and a height of 170 cm [1]

   (ii) a mass of 90 kg and a height of 200 cm [1]

b) (i) Calculate the body mass index for the adults in (a) using the standard equation. [2]

   (ii) Compare the calculated values for body mass index with the estimated values from the nomogram. [1]

c) Explain the significance of the lines on the nomogram for body mass indices of 18.5, 25 and 30. [3]

A2 a) Distinguish between

   (i) monounsaturated and polyunsaturated fatty acids [1]

   (ii) trans and cis unsaturated fatty acids [1]

b) Evaluate the health consequences of diets rich in saturated fatty acids. [3]

A3 Female mammals produce milk for their offspring, containing almost all the nutrients needed by a young mammal.

a) Deduce the types of nutrient contained in milk. [3]

b) Human mothers can either feed their babies on their own milk or on artificial milk.

   Discuss the benefits to babies of feeding on milk from their mother. [3]

c) Cow's milk forms part of the human diet, in some parts of the world.

   Discuss the ethics of consuming cow's milk. [2]

Muscles and fitness

MUSCLES AND MOVEMENT
The information on muscles and movement described on pages 99–100 of Chapter 11 is part of Option B.

FAST AND SLOW MUSCLE FIBRES
Skeletal muscles contain two main types of muscle fibre, fast fibres and slow fibres. Fast muscle fibres are sometimes called Type IIb fibres, and slow fibres are Type I. The differences between the two types of fibre are shown below.

	Fast fibres	**Slow fibres**
Blood supply	Moderate, with some blood capillaries	Excellent, with many blood capillaries
Myoglobin	Little present	Large stores
Mitochondria	Few present	Many present
Cell respiration	Large amounts of the enzymes of glycolysis, giving a high anaerobic capacity	Large amounts of oxidative enzymes in mitochondria, so aerobic capacity is high
Stamina	Low	High
Strength	High	Moderate

Fast muscles fibres contract more rapidly and exert more force per unit of cross-sectional area than slow muscle fibres. Fast fibres can release large amounts of energy for a short period of time by anaerobic respiration, so are useful in high-intensity exercise, for example 100 m sprint races.
Slow fibres release energy more slowly by aerobic cell respiration, but can continue for longer, so are useful in endurance events, for example marathons.
Muscles vary in the proportions of the different types of fibre, both within a person's body and between people. Exercise can affect the proportions within a person's muscles. Moderate-intensity exercise, such as long-distance running or swimming, encourages the development of slow fibres. High-intensity exercise, for example sprinting or weight lifting, encourages the development of fast muscle fibres.
As a result of natural variation and the effects of training programmes, elite athletes vary greatly in the proportions of fast and slow fibres. The figure (below) shows the mean percentage of fast and slow fibres in a thigh muscle in five groups of athletes.

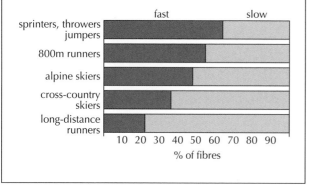

FITNESS
The result of a successful training programme is a condition called **fitness**.

Fitness is the physical condition of the body that allows it to perform exercise of a particular type.

It is important to note that fitness is specific to a particular type of exercise.
During training programmes, it is useful to measure fitness. Various types of measure are used. These often involve measuring speed or stamina.
• **Speed** is the rate at which a movement is performed. The time taken for a movement must be measured. Speed depends mostly on fast muscle fibres. Speed is important in sprinting and football.
• **Stamina** is the ability to continue an exercise for a long time. The maximum duration time is measured. Stamina depends mostly on slow muscle fibres. Stamina is important in rowing and in long-distance running.
Both speed and stamina have their uses as measures of fitness – which is better depends on the type of fitness that is being assessed.

PERFORMANCE ENHANCING SUBSTANCES
Drugs can be used to enhance performances in sport, but there are strong ethical arguments against their use.

1. The long-term health of sportsmen and women who are encouraged to take them may be damaged.
   For example, anabolic steroids can cause men's testes to become smaller and sperm counts to be low. Because anabolic steroids resemble testosterone, they can interfere with women's reproductive system and cause abnormal menstrual cycles. High doses can cause liver disease and there have been reports of athletes who take anabolic steroids suffering from emotional problems, with inappropriately aggressive outbursts. Increased muscle strength allows athletes to generate forces so strong that muscles and tendons can be torn.

2. Drug-users gain an unfair advantage in competitions.
   For example, in Men's 100m finals in a recent Olympic Games, a high proportion of athletes had probably been taking anabolic steroids.

3. Criminals profit from the sale of banned drugs.
   For example, there have been prosecutions of people who have been making substantial profits from the illegal sale of anabolic steroids. If athletes decided not to use these anabolic steroids, these profits could not be made.

There are some ethical arguments in favour of legalizing performance enhancing substances, but few genuine arguments, based on ethics, for their use.

1. Their use might overcome natural variation in physiology, for example, variation in testosterone levels. If all athletes were able to use them, competition might be fairer.

2. If they do enhance performance, spectators might gain more enjoyment from watching sports.

Exercise and cell respiration

GLYCOGEN AND MYOGLOBIN IN MUSCLE

Glycogen is a polysaccharide that is stored in muscle fibres. It is made by linking together glucose molecules and can be broken down to provide a source of glucose for cell respiration. It avoids glucose shortage in muscles during intense or long-duration exercise.

Some muscle fibres contain a red pigment called myoglobin. Oxygen binds to it when the oxygen level in muscle is high. Oxygen is released by myoglobin when the oxygen level in muscle is very low. The role of myoglobin is to act as an oxygen store, allowing muscle fibres to continue aerobic respiration for longer and delaying the formation of lactate.

Causes of muscle fatigue in races

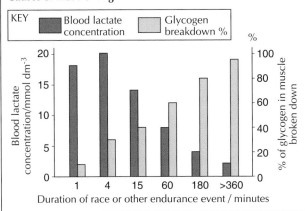

SOURCES OF ATP IN MUSCLES

Muscle contraction requires a supply of energy. It is obtained by converting ATP to ADP. The ADP that is produced must be converted back into ATP, for muscle contraction to continue. There are three ways of doing this:

1. Creatine phosphate

Muscle fibres contain stores of creatine phosphate, which can be used to phosphorylate ADP by this reaction:

creatine + ADP $\longrightarrow$ creatine + ATP
phosphate

This reaction allows ATP to be regenerated for about 8–10 seconds of intense exercise – enough for a 100 m sprint for example. If the duration of exercise is longer then cell respiration must be used.

2. Anaerobic cell respiration

High-intensity exercise, such as sprinting or weight lifting, requires ATP to be supplied at such a rapid rate, that oxygen cannot be supplied fast enough for aerobic cell respiration. Anaerobic cell respiration therefore has to be used. Lactate (lactic acid) is produced by this process and at the same time, hydrogen ions accumulate. Anaerobic respiration can only be used to produce ATP for a maximum of two minutes. Beyond this duration, hydrogen ion concentrations prevent further anaerobic respiration, so high-intensity exercise cannot be continued.

3. Aerobic cell respiration

Oxygen for this type of respiration is brought by blood pumped to the muscle. If oxygen levels in the muscle become low, oxygen supplies can be supplemented for a time by release from myoglobin stores.

Aerobic cell respiration can produce ATP continuously at a rapid enough rate for low-intensity exercise, such as walking or jogging, however long the duration.

EFFECTS OF INCREASING THE INTENSITY OF EXERCISE

As the intensity of exercise increases, the body requires more oxygen for aerobic cell respiration in muscle fibres.

VO_2 is the volume of oxygen that is absorbed by the body per minute and supplied to the tissues.

As the intensity of exercise increases, VO_2 rises, until $VO_{2\,max}$ is reached.

$VO_{2\,max}$ is the maximum rate at which oxygen can be absorbed by the body and supplied to the tissues.

The intensity of exercise can rise above the level where $VO_{2\,max}$ is reached, by using anaerobic respiration. This does not happen as a sudden switch from one type of respiration to the other: as intensity of exercise increases, the percentage of aerobic respiration decreases and the percentage of anaerobic cell respiration increases. Aerobic cell respiration can use fat or carbohydrate as the substrate, but anaerobic cell respiration can only use carbohydrate. For this reason, as the intensity of exercise increases, the use of fat in cell respiration falls and the use of carbohydrate rises until it reaches 100%.

REPAYING THE OXYGEN DEBT

Lactate is carried by blood from muscles to the liver, where it is converted to pyruvate. Oxygen is needed to do this, so if lactate is present in the body there is an **oxygen debt**. If a large amount of lactate builds up during vigorous exercise, a large amount of oxygen is needed to repay the oxygen debt. This is the reason for deep ventilations and a rapid ventilation rate for a time after the exercise.

The pyruvate produced when the oxygen debt is being repaid can either be converted to glucose or can be absorbed by mitochondria and used in aerobic respiration.

CREATINE PHOSPHATE SUPPLEMENTS

Some athletes use creatine as a dietary supplement. An evaluation of its effectiveness is given below.

Question	Answer
Is creatine absorbed from the gut?	Yes.
Can dietary supplementation increase creatine concentrations in muscle?	Yes, but only in athletes with naturally low concentrations. Only small doses of creatine are needed to reach maximal muscle concentrations.
Is the maximum intensity of exercise increased?	There is some evidence of an increase in maximum intensity over short durations.
Can intense exercise be continued for a longer time?	Endurance, involving aerobic cell respiration, is not increased.

Some studies have shown that creatine phosphate supplements cause weight gain by water retention. If this happened, performance might be impaired.

Training and the pulmonary system

MEASURING PULMONARY FUNCTION

The pulmonary system consists of the lungs, the associated muscles and the airways leading to and from the lungs. There are various measures of pulmonary function which are used both during the training of athletes and also in the assessment of patients with diseases of the pulmonary system.

Total lung capacity is the volume of air in the lungs after a maximum inhalation.

Vital capacity is the maximum volume of air that can be exhaled after a maximum inhalation.

Tidal volume is the volume of air that is taken in or out with each inhalation or exhalation.

The term ventilation was defined in Chapter 6 on page 51. Ventilation has a clearer meaning than breathing, so is used in IB Biology.

Ventilation rate is the number of inhalations or exhalations per minute.

EFFECT OF EXERCISE ON VENTILATION

During exercise, increases in ventilation rate and tidal volume usually occur.

- Increases in ventilation rate and tidal volume bring more fresh air to the lungs per minute during exercise.
- This ensures that the concentration of oxygen in air in the alveoli remains high and the concentration of carbon dioxide remains low.
- Blood returning to the lungs during exercise has a higher carbon dioxide concentration and a lower oxygen concentration than at rest.
- Concentration gradients of oxygen and carbon dioxide between alveolar blood and air are therefore steep, maintaining a high rate of gas exchange, with more carbon dioxide diffusing into the alveoli per minute and more oxygen absorbed into the blood, than when the body is at rest.
- This is needed because during exercise the rate of aerobic respiration in muscle fibres increases, with more oxygen used and more carbon dioxide produced per minute.
- When the amount of oxygen supplied per minute to muscles is insufficient, anaerobic respiration has to be used, and the maximum duration of the exercise is not as long as when aerobic respiration is supplying the energy that is needed.

The chart (below) shows the ventilation rate and tidal volume of an athlete running at different speeds.

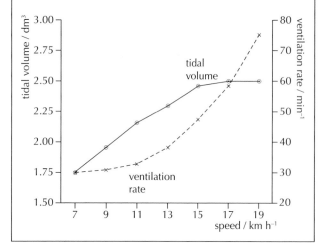

EFFECTS OF TRAINING ON VENTILATION

Training involves repeating exercises that bring the body into the desired state of fitness. There can be effects on the specific muscles used during training and also more general effects on the pulmonary and cardiovascular systems.

Training can reduce the ventilation rate at rest from about 14 to 12 inhalations per minute. This is not because less gas exchange is needed, but because the efficiency of oxygen absorption and carbon dioxide excretion can be increased. Training can increase the maximum ventilation rate from about 40 to 45 inhalations per minute. This is due to strengthening of the muscles used for ventilation.

Training might be expected to increase the vital capacity of the lungs, but if there is any increase, it is only small. Lung capacity appears to be unaffected by training programmes.

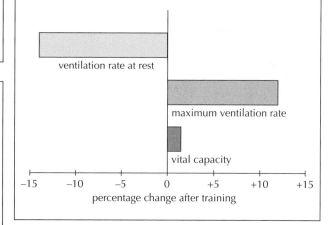

WARM-UP ROUTINES

Most sportsmen and women use warm-up routines to prepare themselves for exercise, whether in a training session or a competitive event. For example, tennis players may do stretching exercises and then spend several minutes hitting the ball across the net gently and practising serves, while not under match pressure.

Various reasons are given to justify warm-up routines:

1. **Improving performance** – blood flow to muscles is increased, supplying more oxygen; muscles become warmer; the rate of respiration can increase, allowing more vigorous and rapid muscle contractions when the competitive event begins.

2. **Psychological preparation** – if a specific warm-up routine is used every time before an event, it may help to get the body mentally ready for physical activity and for competition, by adrenalin secretion or other means.

3. **Preventing injuries** – muscles that have been warmed up and tendons and ligaments that have been gently stretched may be less vulnerable to injuries.

The evidence for the effectiveness of warm-up routines is rather thin and is based mostly on small numbers of individual cases (anecdotal evidence) rather than on controlled trials with large numbers. Athletes are understandably reluctant to compete without warming up, for research purposes. Some anecdotal evidence suggests that warming up may not be essential – reserves often compete successfully in matches, after little or no warming up!

Training and the cardiovascular system

MEASURING HEART FUNCTION

The cardiovascular system consists of the blood, the heart and the blood vessels. Heart function can be assessed using these measures:

Heart rate is the number of contractions of the heart per minute.

Stroke volume is the volume of blood pumped out with each contraction of the heart.

Cardiac output is the volume of blood pumped out by the heart per minute.

Venous return is the volume of blood returning to the heart via the veins per minute.

EFFECTS OF TRAINING ON THE HEART

Training can increase the thickness of the heart wall and the volume of the ventricles. The stroke volume is therefore larger, both at rest and during exercise.

The body does not need a larger cardiac output at rest, so the heart rate can be lower. Training can reduce the heart rate at rest to 50 beats per minute.

At any level of intensity of exercise, the heart rate is lower after training, because of the larger stroke volume.

The maximum heart rate is not greatly affected by training, but because of the greater stroke volume, cardiac output is much greater at maximum heart rate after training. This allows the trained athlete to perform a much greater intensity of exercise.

EFFECTS OF EXERCISE ON THE CARDIOVASCULAR SYSTEM

1. Venous return increases during exercise. When muscles in the legs and arms contract, the muscles become shorter and wider and so exert pressure on adjacent veins. There are valves in these veins, ensuring that blood flows towards the heart. Pressure therefore causes blood to be squeezed along veins to the heart, increasing venous return. This allows cardiac output to be increased.

2. Cardiac output increases as a result of increases in heart rate and stroke volume. Exercise involves a rise in carbon dioxide production by muscles. Absorption of this extra carbon dioxide into the blood causes a decrease in blood pH. The brain detects the pH decrease and sends impulses to the heart's pacemaker, causing the increase in cardiac output.

3. The distribution of blood changes when exercise starts. Arterioles supplying the organs of the body can narrow or widen, decreasing or increasing the flow of blood. The lists below show which organs receive more blood during exercise than at rest, which receive less blood and which receives the same volume.

More during exercise	Less during exercise	Same volume
skeletal muscles	kidneys	brain
heart wall	stomach	
skin	intestines	

In summary, during exercise, blood returns to the heart and is pumped out at a greater rate. Much of this blood flows to the muscles, increasing the supply of oxygen, allowing an increase in the rate of aerobic respiration and ATP supply for muscle contraction.

RISKS AND BENEFITS OF EPO

Athletes sometimes increase the amount of red blood cells, as a proportion of the volume of their blood. This is called the packed cell volume (PCV). At sea level a normal PCV is 0.4–0.5. There are several ways of increasing PCV above 0.5, including the following:

1. Injections of EPO (erythropoietin), a naturally produced hormone that stimulates red blood cell production.

2. Blood transfusions, shortly before an event. Often the transfused blood was removed from the athlete's body long enough before the event for the blood cells to have been replaced.

There are clear benefits in terms of performance of increasing PCV. As these cells transport oxygen, the larger the numbers of them, the greater the rate at which oxygen can be carried around the body by the blood. With more oxygen, skeletal muscles can contract more vigorously.

There are also some risks. High levels of PCV increase the chance of blood clot formation (thrombosis). Blood clots cause heart attacks and strokes. There have been deaths among cyclists and other athletes, who had used one or other of the methods above to increase PCV.

INJURIES TO MUSCLES AND JOINTS

Vigorous exercise sometimes causes injuries to muscles and joints.

- **Torn muscles** – excessive stretching causes muscle fibres, or more rarely an entire muscle, to tear, for example the quadriceps or hamstrings.

- **Sprains** – abnormal movement at a joint causes stretching or minor tearing of ligaments, for example joints in the fingers or the ankle (see figure, left).

- **Torn ligaments** – large abnormal movements cause ligaments to tear completely, for example the cruciate ligaments in the knee.

- **Dislocation** – abnormal movement at a joint causes the bones to move out of alignment. Usually ligaments will be torn at the same time.

- **Intervertebral disc damage** – abnormal movements or heavy loads cause the soft centre of a disc to bulge out, through a tear in the disc wall (see question 3 on page 120).

Ankle sprains

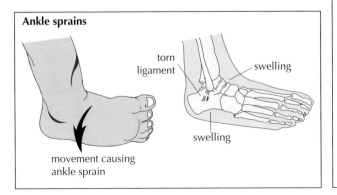

torn ligament

swelling

swelling

movement causing ankle sprain

EXAM QUESTIONS ON OPTION B – PHYSIOLOGY OF EXERCISE

B1 Humans and other mammals can store oxygen in the lungs, in muscles and in the blood. The pie charts below show the volume of oxygen (cm^3) per kilogram of body mass stored in these tissues in humans and in a marine mammal, the Weddell seal.

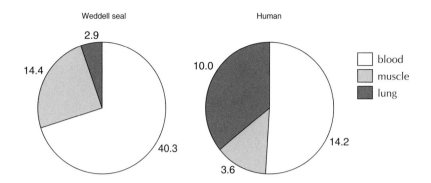

a) Compare the total amount of oxygen stored per kilogram of body mass in seals with that in humans. [1]

b) Compare the proportions of oxygen stored in blood, muscle and lung of seals with those in humans. **(No calculations are required)**. [3]

c) Suggest **three** factors which affect how much oxygen can be stored in muscle in the body of a mammal. [3]

B2 a) Draw a diagram to show the structure of a sarcomere. [3]

b) Outline the role of ATP in muscle contraction. [3]

c) Compare cardiac output at rest and when vigorous muscle contractions are being performed. [1]

B3 The scan (right) shows damage to intervertebral discs in the neck of a person. Grey and white matter in the spinal cord can be distinguished.

a) State the number of discs that are damaged. [1]

b) Describe the damage to these discs. [2]

c) State the other part of the person's body that is affected by the disc damage. [1]

d) Suggest how damage to intervertebral discs may be caused. [2]

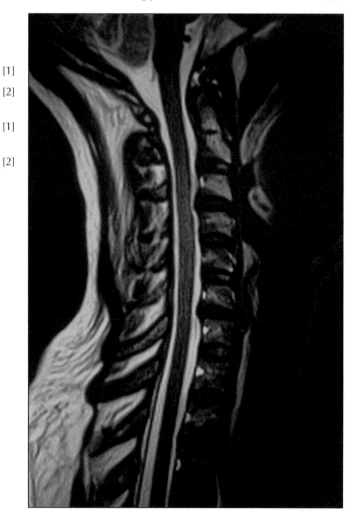

EXAM QUESTIONS ON OPTION C – CELLS AND ENERGY

Topics in Option C are covered on pages 66–81.

C1 The rate of photosynthesis in plants can be influenced by many factors. Experiments were carried out to investigate the effect of high and low light intensities on photosynthesis at different temperatures. All other factors were kept constant. A summary of the results is presented in the graph below.

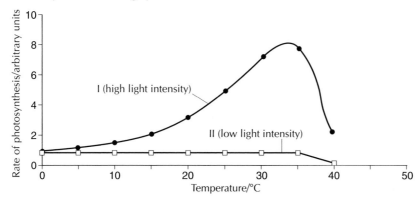

a) State the name of one limiting factor of photosynthesis, apart from temperature and light intensity. [1]

b) (i) Deduce the factor limiting the rate of photosynthesis in experiment I, between 0 and 30°C. Give a reason for your answer. [2]

   (ii) Discuss which factor limits the rate of photosynthesis in experiment I, between 35 and 40°C. [2]

c) Suggest one explanation for the difference between the results of experiments I and II. [2]

C2 Enzymes can be inhibited competitively and non-competitively.

a) State one example of:

   (i) a competitive inhibitor [1]

   (ii) a non-competitive inhibitor [1]

b) Compare competitive and non-competitive inhibition by stating one similarity and one difference in the table below. [2]

	Competitive inhibition	Non-competitive inhibition
Similarity		
Difference		

C3 The reactions of part of aerobic cell respiration are shown below.

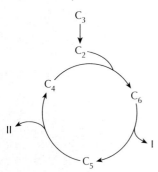

a) Identify the compounds C_3 and C_2. [2]

b) Identify I and II. [2]

c) State one other product of these reactions. [1]

d) State the name of the cycle of reactions. [1]

Origin of life on earth

SPONTANEOUS ORIGIN OF LIFE

Pasteur showed in an experiment in the 19th century that spontaneous generation of life from inorganic matter does not now take place – cells can only be formed from other cells. This is not surprising, as even the simplest prokaryotic cells are very complicated. Nevertheless, when the Earth was first formed there were no living cells on it, so at some stage the first living cells must have appeared. Claims that this happened 3.8 billion years ago are now disputed and the oldest undisputed bacterial fossils are in the Gunflint cherts of Ontario, dating from 1.9 billion years ago.

Four processes would have been needed for the first cells to form:

• chemical reactions to produce simple organic molecules, such as amino acids, from inorganic molecules, such as water, carbon dioxide and ammonia
• assembling of these simple organic molecules into polymers, for example, polypeptides from amino acids
• formation of polymers that can self-replicate – this allows inheritance of characteristics
• development of membranes, to form spherical droplets, with an internal chemistry different from the surroundings, including the polymers that held the genetic information.

The product of these four processes would have been cell-like structures. Natural selection could have operated on them, allowing evolution to begin.

ORIGIN OF ORGANIC COMPOUNDS

Various possible locations have been suggested for the synthesis of the organic compounds needed for the origin of life.

1. Miller and Urey's experiments suggest that organic compounds could have been synthesized by chemical reactions in the atmosphere and in water, on the surface of the Earth.

2. There are hydrothermal vents deep in the oceans, with chemicals welling up from the rocks below. Around these vents, there are very unusual chemical conditions, which might have allowed the spontaneous synthesis of the organic compounds from which the first organisms evolved.

3. Some theories involve an extraterrestrial origin for organic compounds. Experiments by scientists working with NASA have shown that organic compounds and proto-cells could have formed in cold interstellar space. They might then have been delivered to the Earth by meteorites, comets or interplanetary dust. There was a heavy bombardment of the Earth by meteorites 4000 million years ago, which might have brought the organic compounds that became organized into the first living organisms.

MILLER AND UREY'S EXPERIMENTS

In 1953, Stanley Miller and Harold Urey investigated the theory that organic compounds could have formed spontaneously on Earth. They recreated the conditions that probably existed on Earth before living organisms were present. Inside their apparatus (below) they mixed the gases ammonia, methane and hydrogen to form a reducing atmosphere. Electrical discharges and the boiling and condensing of water simulated lightning and rainfall. After one week, the clear water in the apparatus had turned to a murky brown. Analysis revealed many organic compounds, including fifteen amino acids. Miller and Urey concluded that organic compounds could have formed spontaneously on Earth, before there were any living organisms here.

Miller and Urey's apparatus

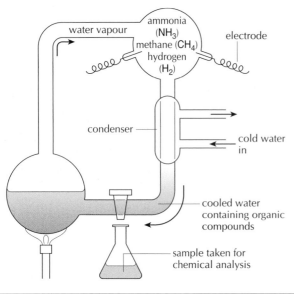

THE ROLE OF RNA IN THE ORIGIN OF LIFE

In modern prokaryotes, the various parts of the genetic mechanism cannot function without each other. For example, genes cannot be replicated without enzymes and enzymes cannot be made without genes. It seems inconceivable that the whole mechanism could have evolved at once, but gradual evolution would have required simpler intermediate stages. One possibility is the use of RNA instead of both DNA and enzymes. RNA may have had a very significant role in the origin of life. It has two properties that would have allowed it to do this – **catalysis** and **self-replication**.

1. RNA catalyses a broad range of chemical reactions. It could therefore have taken the role that is carried out by proteins (enzymes) in the organisms that now exist on Earth.
RNA still catalyses some reactions, for example peptide bond formation during protein synthesis in the ribosome.

2. RNA is capable of self-replication – one molecule can form a template for the production of another molecule, following the rules of complementary base pairing. If the newly synthesized molecule is then used as a template, a replicate of the original molecule will be produced.

No biological mechanisms now exist for self-replication by RNA molecules, but this is not surprising as RNA was superseded, billions of years ago, by DNA as the genetic material and by proteins as the catalysts of life. There are various reasons for the DNA–protein world replacing the RNA world. One possibility is that the maximum length of RNA molecules is about 1500 nucleotides – this places a severe restriction on the amount of genetic information that can be held. RNA viruses, for example, have a very small genome.

Origin of prokaryotes and eukaryotes

MEMBRANES AND PROTOBIONTS

To form the first cells, membranes were needed to separate cytoplasm and its metabolism from the surrounding fluid. Phospholipids naturally group together to form bilayers in water. These bilayers form spherical structures enclosing a droplet of fluid, similar to the vesicles that are now found in cells.

Water containing these membrane-bound **microspheres** is called **coacervate** and is viscous and cloudy in appearance.

Because of their hydrophobic properties, bilayers of phospholipid would have allowed an internal environment to develop, different from the surrounding environment.

These primitive cell-like structures, that may have preceded living cells, are called **protobionts**. To become cells, they would have had to develop genetic mechanisms to allow reproduction and the transmission of characteristics to offspring. The details of this transition are not yet understood.

PROKARYOTES AND THE ATMOSPHERE

The first organisms on Earth to use photosynthesis for the synthesis of organic compounds were prokaryotes. When these organisms started to use water as source of hydrogen in photosynthesis, oxygen started being released as a waste product into the atmosphere. There is evidence that before this time there was little oxygen in the atmosphere.

Concentrations of oxygen built up over a relatively short period – about a hundred million years! This was probably due to the activity of photosynthetic prokaryotes. Other prokaryotic organisms were able to use aerobic cell respiration, once the atmosphere contained oxygen.

Rocks in Greenland dating from 3.7–3.8 million ago, called the banded iron formation, give evidence of oxygen in the atmosphere. This suggests that prokaryotic cells had evolved and were producing oxygen by then. Among existing organisms, photosynthetic bacteria in hot springs and other extreme environments are probably most similar to these early prokaryotes.

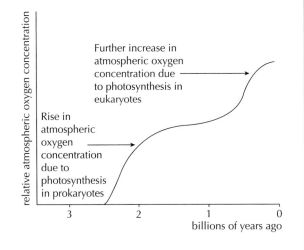

THE ENDOSYMBIOTIC THEORY

Eukaryotic cells contain mitochondria and chloroplasts, which are not found in prokaryotic cells. If eukaryotic cells evolved from prokaryotic cells, the origin of these organelles must be explained.

According to the **endosymbiotic theory**, both mitochondria and chloroplasts have evolved from independent prokaryotic cells, which were taken into a larger heterotrophic cell by endocytosis. Instead of being digested, the cells were kept alive and continued to carry out aerobic respiration and photosynthesis. The characteristics of mitochondria and chloroplasts support the endosymbiotic theory.

- They grow and divide like cells.
- They have a naked loop of DNA, like prokaryotes.
- They synthesize some of their own proteins using 70S ribosomes, like prokaryotes.
- They have double membranes, as expected when cells are taken into a vesicle by endocytosis.

Some biologists have suggested that flagella and cilia also have an endosymbiotic origin, but the evidence for this is less clear. The evolution of eukaryotes from prokaryotes did not just involve the development of mitochondria, chloroplasts and possibly cilia and flagella. Eukaryotic chromosomes, meiosis and sexual reproduction also had to evolve. Once this had happened, evolution could take place at a much more rapid pace than before and there was what has sometimes been described as an explosion of life on Earth.

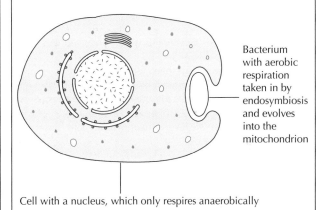

Bacterium with aerobic respiration taken in by endosymbiosis and evolves into the mitochondrion

Cell with a nucleus, which only respires anaerobically

GENE POOLS AND ALLELE FREQUENCIES

A new individual, produced by sexual reproduction, inherits genes from its two parents. If there is random mating, any two individuals in an interbreeding population could be the two parents, so the individual could inherit any of the genes in the interbreeding population. These genes are called the **gene pool**.

A gene pool is all the genes in an interbreeding population.

Many genes have different alleles. In a typical interbreeding population, some alleles will be commoner than others. How common an allele is can be assessed using **allele frequency**.

Allele frequency is the frequency of an allele, as a proportion of all alleles of the gene in the population.

Allele frequency can range from 0.0 to 1.0.
Evolution always involves a change in allele frequency in a population's gene pool, over a number of generations.

Species and speciation

WHAT IS A SPECIES?

Biologists have been arguing about the exact meaning of the term species for over two hundred years. Before the discovery that species can evolve, a species was regarded as a type of living organism with fixed characteristics, which distinguish it from other species. This is known as the morphological definition of a species. It is still a useful idea. Species can usually be distinguished from each other by their characteristics – this is how specimens are identified.

However the morphological definition does not recognize the fact that species evolve. If two populations with similar but not identical characteristics are geographically separated, they may be in the gradual process of splitting from one species into two separate ones. It is not easy for a taxonomist to decide whether to classify them as one or two species and some criterion is needed to decide. The reason for members of a species having common features is that they interbreed with each other. The reason for the characteristics of one species being different from those of another is that the two species do not interbreed and are evolving separately. Biologists now regard interbreeding as a more important criterion than morphology. The biological definition of a species is *a group of actually or potentially interbreeding populations, with a common gene pool, which are reproductively isolated from other such groups.*

Only if two separated populations can be shown to be capable of interbreeding should they be classified as one species.

The biological species definition is widely accepted, but it does cause some problems.
- Many sibling species have been found. These are species that cannot interbreed, but show no significant differences in appearance. Although separate species, they are very difficult for ecologists to identify. For example the Pipistrelle bat in Britain was recently shown to be two sibling species.
- Some pairs of species that are clearly different in their characteristics will interbreed. Many plant species hybridize and some animals also, e.g. ruddy ducks and white-headed ducks (below).
- Some species always reproduce asexually, so the members of a population do not interbreed. The biological species definition is therefore unusable.
- Fossils cannot be classified according to the biological species definition, as it is impossible to decide with which organisms they would have been able to interbreed.

Two animal species that can interbreed

Ruddy duck   White-headed duck

SPECIATION

The formation of new species is called speciation. New species are formed when a pre-existing species splits. This usually involves the isolation of a population from the remainder of its species and thus the isolation of its gene pool. The isolated population will gradually diverge from the rest of the species if natural selection acts differently on it. Eventually the isolated population will be unable to interbreed with the rest of the species – it has become a new species. Speciation can either be **allopatric** or **sympatric**.

1. Allopatric speciation occurs when members of a species migrate to a new area, forming a population that is geographically isolated from the rest of the species. Interbreeding is impossible – **geographical isolation** acts as a barrier between the gene pools of the populations. The populations can therefore split to form separate species. This can happen repeatedly, for example with the lava lizards of the Galápagos.

Distribution of lava lizards on the Galapagos Islands

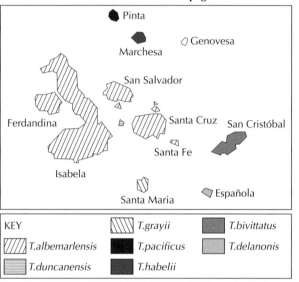

KEY			
	T.grayii		T.bivittatus
T.albemarlensis	T.pacificus		T.delanonis
T.duncanensis	T.habelii		

2. Sympatric speciation occurs when two varieties of a species live in the same geographical area, but do not interbreed. Two examples will be considered here:

(a) The apple maggot fly (*Rhagoletis pomonella*) of North America is an example. It used to lay its eggs only on hawthorn fruits, which were the food of its larvae. It now also infests non-native apple trees as well. One strain of this species now lays its eggs on apple fruits and other strains on hawthorn fruits. Because the fruits ripen at different times, adults of the two strains emerge and mate at different times, so there is a **behavioural** or **temporal** barrier between the gene pools. There are differences in allele frequencies of the two strains showing that sympatric speciation has started to occur.

(b) Barriers between gene pools can also occur by **hybrid infertility**, often due to polyploidy. If there are some tetraploid individuals in a population, the gametes that they produce will be diploid. Hybrids produced when diploids mate with tetraploids will be triploid. These hybrids will always be sterile as meiosis fails in triploid cells. So, diploids can only produce fertile offspring by mating with diploids and tetraploids by mating with other tetraploids. Plants in the genus *Rumex* are good examples of speciation by polyploidy. The basic diploid chromosome in this genus is 20, but *Rumex obtusifolius* has 40 chromosomes and so is a tetraploid. *Rumex crispus* is a hexaploid with 60 chromosomes and there is even a decaploid species – *Rumex hydrolapathum*, with 200 chromosomes!

Trends in evolution

ADAPTIVE RADIATION

Speciation often happens repeatedly, to form a group of species from one ancestral species. Sometimes each species then evolves in very different ways. This is called **divergent evolution**. By becoming adapted to different ecological roles, the different species avoid competition with each other. If species in a group diverge rapidly in this way, it is called **adaptive radiation**. This can happen when the group has a characteristic that gives it a competitive advantage over existing species or where there are opportunities that no other species are utilizing. Darwin's finches on the Galápagos archipelago are an example. Mammals are another group that demonstrate adaptive radiation. The figure (below) shows examples of the mammalian pentadactyl limb, derived from one ancestral mammal.

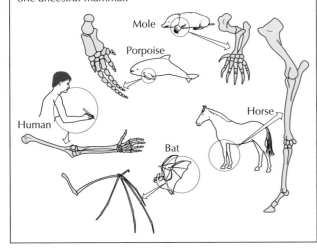

CONVERGENT EVOLUTION

Living organisms often find the same solutions to particular physiological problems. If natural selection acts in the same way, in different parts of the world, species can become remarkably similar, despite not being closely related. This is called **convergent evolution**. It is the converse, in many ways, of adaptive radiation. Instead of closely-related species showing striking differences, unrelated species show striking similarities. Cacti and euphorbias are examples of this. The photographs below show a cactus from the south-west USA and a euphorbia from Madagascar.

Ocotillo (a cactus) from south-west USA

Allaudia (a euphorbia) from Madagascar

RATES OF EVOLUTION

There has been much discussion among biologists about rates of evolution. One idea, called **gradualism**, is that evolution proceeds very slowly, but large changes can gradually take place over long periods of time.

This does not fit in with the fossil record, which shows periods of stability, with fossils showing little evolution, followed by periods of sudden major change. The periods of stability may be due to equilibrium where living organisms have become well adapted to their environment so natural selection acts to maintain their characteristics. The periods of sudden change that occasionally occur, may correspond with rapid environmental change, caused for example by volcanic eruptions or meteor impacts. New adaptations would be necessary to cope with changed environmental conditions, hence strong directional selection and rapid evolution. This view of the pace of evolution is called **punctuated equilibrium**.

TRANSIENT POLYMORPHISMS

A population in which there are two alleles of a gene in the gene pool is **polymorphic**. If one allele is gradually replacing the other the population shows **transient polymorphism**. The peppered moth, *Biston betularia,* is an example of this. In both Britain and the United States, melanic forms were discovered in the 19th century (*carbonaria* and *swettaria*). Both of these forms are due to dominant alleles of a gene that affects wing colour. These dominant alleles increased in frequency in some areas, where air pollution caused natural selection to favour moths with dark wings.

In many areas the dominant alleles then decreased in frequency in the second half of the 20th century. This was because there had been control of air pollution and the cleaner air meant that natural selection favoured the lighter coloured moths. The dominant alleles for darker wings will probably reduce to very low frequencies in areas where there is clean air.

BALANCED POLYMORPHISMS

Sometimes two alleles of gene can persist indefinitely in the gene pool of a population. It is not therefore a transient polymorphism and instead is called **balanced polymorphism**. The most thoroughly researched example of a balanced polymorphism is sickle cell anemia (see page 23).

- Individuals with the genotype Hb^AHb^A do not develop sickle cell anemia but are susceptible to malaria.
- Individuals with the genotype Hb^SHb^S are resistant to malaria, but develop severe sickle cell anemia.
- Heterozygous individuals (Hb^AHb^S) do not develop sickle cell anemia and are resistant to malaria. They are therefore the best adapted in areas where malaria is found.

Both of the alleles of the hemoglobin gene therefore tend to persist in malarial areas. The sickle cell allele has increased in frequency to high levels in some of these areas. In parts of Africa, as many as 40% of the population are carriers of the sickle cell allele, so show resistance to malaria.

Human origins

HUMANS AS PRIMATES

The primates are an order of mammals, including apes, monkeys, tarsiers and lemurs. They were given this name because they were considered to be the highest order of animals. Humans are classified as primates, because they have the anatomical features that are characteristic of this order:

- Grasping limbs, with long fingers and a separated opposable thumb.
- Mobile arms, with shoulder joints allowing movement in three planes and the bones of the shoulder girdle allowing weight to be transferred via the arms.
- Stereoscopic vision, with forward facing eyes on a flattened face, giving overlapping fields of view.
- Skull modified for upright posture.

The unavoidable conclusion, so shocking when it was first drawn, is that humans evolved from other primate species. There has been a huge research effort to try to find out how this occurred.

TRENDS IN HOMINID FOSSILS

Hominids are members of the family Hominidae – the family that includes humans. A notable feature of this family is walking on two legs – bipedalism.
Homo sapiens is currently the only species of hominid but other species existed in the past. At various stages in hominid evolution, several species almost certainly co-existed, for example *Homo sapiens* with *Homo neanderthalensis*. Many hominid fossils have been found, dated, and assigned to a species. These fossils show evolutionary trends:
- including increasing adaptation to bipedalism
- increasing brain size in relation to body size.

Brain sizes of *Homo* and *Australopithecus*

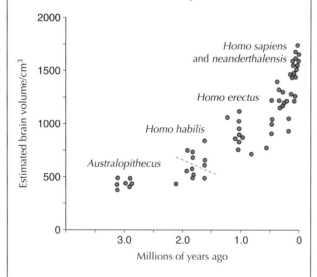

Other trends are shown in the figure (right).
Fossils of *Ardipithecus* were found in Ethiopia, *Australopithecus* and *Homo habilis* fossils were all found in Southern or Eastern Africa. *Homo erectus* fossils were found in Eastern Africa, but also in Asia, indicating that there was migration out of Africa. *Homo neanderthalensis* fossils were found in Europe and *Homo sapiens* in many parts of the world indicating further migrations.

Ardipithecus ramidus (4.4 million years)
Only fragments of skulls and other bones have been found so far. They suggest characters intermediate between chimpanzees and *Australopithecus*:
- small numbers of large molars, like chimps
- incisors slightly smaller than those of chimps
- canines blunt and projecting less than those of chimps
- foramen magnum (hole through which spinal cord enters the skull) further forward than in apes, suggesting *Ardipithecus* was at least partially bipedal.

Australopithecus afarensis (4 to 2.5 million years)

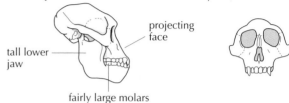

projecting face
tall lower jaw
fairly large molars

Australopithecus africanus (3 to <2.5 million years)

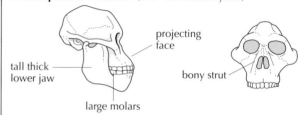

projecting face
tall thick lower jaw
bony strut
large molars

Homo habilis (2.4 to 1.6 million years)

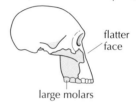

flatter face
large molars

Homo erectus (1.7 to 1.8 million years)

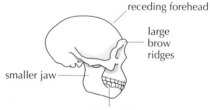

receding forehead
large brow ridges
smaller jaw
smaller molars

Homo neanderthalensis (500 000 years)

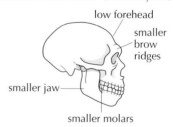

low forehead
smaller brow ridges
smaller jaw
smaller molars

Homo sapiens (140 000 to 70 000 years)

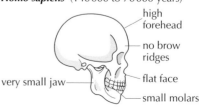

high forehead
no brow ridges
flat face
very small jaw
small molars

Human evolution

TRACING HUMAN EVOLUTION

Our understanding of human evolution is based mostly on fossils and the hominid fossil record contains many gaps. It is not unusual for the fossil record of a group of organisms to be incomplete. Only a tiny proportion of animal bodies become fossilized. It is far more usual for animal bodies to be eaten by detritivores, decomposed by bacteria or broken down chemically. Organic acids in decomposing material react with alkali in bones and teeth, for example. Hominid fossils consist only of bones and teeth. These remains have been preserved where dry sediments have quickly covered them and have remained undisturbed.

Because the hominid fossil record is incomplete, it is far from clear how the different species of hominid are related. Many details of human evolutionary origins are also uncertain. Discoveries of small numbers of fossils can cause major changes in the prevailing theories. For example, there have been recent finds of an *Australopithecus* species, with characteristics between those of *Ardipithecus ramidus* and *Australopithecus afarensis*. Dating of fossils of the three species from the Afar district of Ethiopia suggests that they did not co-exist, but instead they form an evolutionary lineage. As with most theories about human evolution, this has been disputed!

DATING FOSSILS

To place fossils into a sequence it is necessary to know their dates. Fossils, or the rocks containing fossils, can be dated using radioisotopes – radioactive isotopes of chemical elements. When an atom of a radioisotope decays, it changes into another isotope and gives off radiation. The rate of decay varies between different radioisotopes and is expressed as the **half-life**.

The half-life is the time taken for the radioactivity to fall to half of its original level.

The graph below shows a decay curve for radioisotopes.

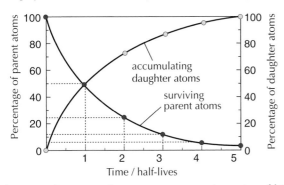

The two radioisotopes that are most commonly used are ^{14}C and ^{40}K. In radiocarbon dating the percentage of surviving ^{14}C atoms in the sample is measured. In potassium-argon dating, the proportions of parent ^{40}K atoms and daughter ^{40}Ar atoms are measured. In both methods the age in half-lives can then be deduced from the decay curve. The half-life of ^{14}C is 5730 years so it is useful for dating samples that are between one thousand and one hundred thousand years old. The half-life of ^{40}K is 1250 million years so it is useful for dating samples older than 100 000 years.

HOMINID DIETS AND BRAIN SIZE

The brains of early hominids (*Australopithecus*) were only slightly larger in relation to body size than the brains of apes. The powerful jaws and teeth of *Australopithecus* indicate a mainly vegetarian diet.

About 2.5 million years ago Africa became much cooler and drier. Savannah grassland replaced forest. This change of habitat may have prompted the evolution of the first species of *Homo*, with the development of increasingly sophisticated tools and a change to a diet that included meat obtained by hunting and killing large animals.

This change in diet corresponds with the start of the increase in brain size of hominids. This was due to continued rapid brain growth after birth. In apes and earlier hominids, brain growth slows after birth. The correlation between the change in diet and the increases in brain size can be explained in two ways:

1. Eating meat increases the supply of protein, fat and energy in the diet, making it possible for the growth of larger brains.

2. Catching and killing prey on the savannas is more difficult than gathering plant foods, so natural selection will have favoured hominids with larger brains and greater intelligence.

GENETIC AND CULTURAL EVOLUTION

The large brains of *Homo sapiens* and other species of *Homo* allow much to be learned, both during the long period of childhood and during adulthood.
Language, tool making skills, hunting techniques, methods of agriculture, religion, art and many other forms of behaviour are passed on from one generation of a tribe or other group to the next by teaching and learning. These things are the culture of the group.
New methods, inventions or customs can be incorporated into what is passed on. This is called **cultural evolution** and is different from the type of evolution that involves natural selection between inherited differences – **genetic evolution**.

- Cultural evolution does not involve changes in allele frequencies in the gene pool.
- Changes due to cultural evolution can happen during one human lifetime, whereas genetic evolution happens over generations, so cultural can be much more rapid than genetic evolution.
- Cultural evolution involves characteristics acquired during a person's life (nurture) whereas genetic evolution involves characteristics that are inherited (nature).

In the recent evolution of humans, cultural evolution has been very important and has been responsible for most of the changes in the lives of humans over the last few thousand years. This is much too short a period for genetic evolution to cause much change. Also some aspects of cultural evolution, for example the development of medicine, have reduced natural selection between different genetic types and therefore genetic evolution.

The Hardy–Weinberg principle

THE HARDY–WEINBERG EQUATION

Evolution involves changes in allele frequency, so it is a useful skill to be able to do calculations involving allele frequencies. The Hardy–Weinberg equation is often used for this.

If there are two alleles of a gene in a population, the frequency of the alleles in the population is usually represented by the letters p and q. The total frequency of the alleles in the population is 1.0, so

$$p + q = 1.$$

If there is random mating in a population, the chance of inheriting two copies of the first of the two alleles is $p \times p$. The chance of inheriting two copies of the second of the two alleles is $q \times q$. The expected frequency of the two homozygous genotypes is therefore p^2 and q^2. The expected frequency of the heterozygous genotype is $2pq$. The sum of all of these frequencies is 1.

$$p^2 + 2pq + q^2 = 1$$

This is called the **Hardy–Weinberg equation**. It is represented by the figure below.

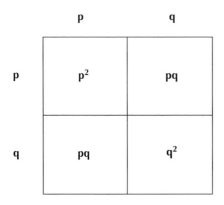

Occasionally it is possible to test whether the proportions of genotypes in a population fit this equation. Both allele frequencies and genotype frequencies must be known. The table below shows the results of a survey of the MN blood group gene in a Japanese town. The two alleles of this gene are co-dominant.

Allele frequencies in the parental generation

M allele: $p = 0.525$ N allele: $q = 0.475$

Genotype frequencies in the offspring

	Predicted	Actual
MM	$p^2 = 0.276$	0.274
MN	$2pq = 0.499$	0.502
NN	$q^2 = 0.225$	0.224

The results of the survey show that the actual genotypes fit those predicted by the Hardy–Weinberg equation very closely. The population therefore follows the **Hardy–Weinberg Principle**.

HARDY–WEINBERG CALCULATIONS

The Hardy–Weinberg equation can be used in calculations if certain assumptions can be made:
- that there is random mating in the population
- natural selection does not cause higher mortality of individuals with one allele than the other
- there is no mutation
- the population is not very small
- there is no immigration or emigration.

If these assumptions are correct, the population is said to be in Hardy–Weinberg equilibrium and the equation is valid for that population.

1. **Example of calculating phenotype frequencies**
   An experimental plot of pea plants is established by sowing seeds of pure breeding tall and dwarf varieties, in a ratio of three tall to one dwarf. The plants are allowed to produce and disperse seeds naturally.
   The conditions on the plot ensure that all the assumptions for the Hardy–Weinberg equation are satisfied. What proportions of tall and dwarf pea plants are expected after several generations?

   Genotypes of pure-breeding varieties are TT and tt
   Frequency of T (p) is 0.75
   Frequency of t (q) is 0.25
   Frequency of dwarf plants $(tt) = q^2 = 0.25^2 = 0.0625$
   Frequency of tall plants $= 1 - q^2 = 1 - 0.0625 = 0.9375$
   This frequency could also be calculated as $p^2 + 2pq$

2. **Example of calculating allele frequencies**
   The gene that controls the ability to taste phenylthiocarbamide (PTC) has two alleles.

   Ability to taste PTC is due to the dominant allele (T) and non-tasting is due to the recessive allele (t).
   1600 people were tested in a survey.
   461 were non-tasters – a frequency of 0.288.
   Their genotype was homozygous recessive ($t\,t$).
   If q = frequency of t allele, $q^2 = 0.288$ so $q = 0.537$
   If p = frequency of T allele, $p = (1 - q) = 0.463$

3. **Example of calculating genotype frequencies**
   Cystic fibrosis is a genetic disease caused by recessive alleles of a chloride channel gene.

   More than 27000 people in Scotland, none of whom had cystic fibrosis, were screened to see if they were carriers of an allele for cystic fibrosis.

   From the frequency of carriers, the allele frequencies in the population could be deduced:

   Frequency of normal allele $= p = 0.9776$
   Frequency of cystic fibrosis allele $= q = 0.0224$

   When these people have children, the chance of their child being homozygous for the cystic fibrosis allele is

   $$q^2 = (0.0224)^2 = 0.000502$$

   This is equivalent to about one child in 1900 with cystic fibrosis.

   The chance of their child being a carrier is

   $$2pq = 2(0.9776 \times 0.0224) = 0.0438$$

   This is equivalent to about one child in 23 being a carrier.

REASONS FOR CLASSIFICATION

Classification in biology is arranging living organisms into groups. There are many advantages:

- **Species identification** – it is easier to find out to which species an organism belongs with organisms classified rather than forming a disorganized catalogue.
- **Predictive value** – if several members of a group have a characteristic, another species in this group will probably also have this characteristic.
- **Evolutionary links** – species that are in the same group probably share characteristics because they have evolved from a common ancestor, so the classification of groups can be used to predict how they evolved.

These are advantages of a natural classification – one that matches the evolutionary origins of the species in the group. Artificial classification systems sometimes help with species identification, but have no other value. An example of an artificial classification is putting insects, birds and bats into one group because they fly. The wings of these animals are examples of **analogous structures** – structures with a common function, but a different evolutionary origin. Natural classification is based on **homologous structures** – structures that have a common evolutionary origin, even if their function is different. The pentadactyl limb (see page 125) is an examples of a homologous structure in mammals. Organisms with homologous structures should be classified in the same group because they must have common ancestry, even if they look superficially different.

BIOCHEMISTRY AND COMMON ANCESTRY

There are remarkable similarities between living organisms in their biochemistry.

- All use DNA (or RNA) as their genetic material.
- All use the same universal genetic code, with only a few insignificant variations.
- All use the same 20 amino acids in their proteins.
- All use left, and not right-handed amino acids.

The similarities in amino acid composition are striking because many other amino acids, in both left and right-handed versions, were available when life evolved, according to Miller and Urey's experiments.

These biochemical similarities suggest very strongly that all organisms have evolved from a common ancestor, which had all of these characteristics.

PHYLOGENY AND BIOCHEMISTRY

Tracing evolutionary links and origins is called **phylogeny**. The phylogeny of many groups has been studied by comparing the structure of a protein or other biochemical that they contain. Usually the results match the existing classification of the group. The diagram (above right) shows the results of a study based on DNA sequences.

Chimpanzees and gorillas are currently in a family with orang-utans, but should probably be placed in the same family as humans, according to this DNA evidence.

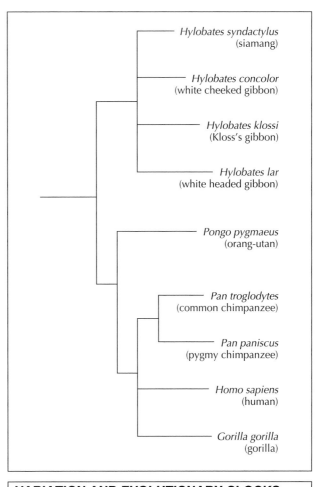

VARIATION AND EVOLUTIONARY CLOCKS

Differences in the base sequence of DNA and therefore in the amino acid sequence of proteins, accumulate gradually over long periods of time. There is evidence that differences accumulate at a roughly constant rate. They can therefore be used as an evolutionary clock. The number of differences in amino acid sequence can be used to deduce how long ago species split from a common ancestor.

For example, mitochondrial DNA from three humans and four related primates has been completely sequenced. From the differences in base sequence, a hypothetical phylogeny has been constructed, shown (below). Using the numbers of differences in base sequence as an evolutionary clock, these approximate dates for splits between groups have been deduced:

70 000 years ago, Europeans–Japanese split
140 000 years ago, African–European/Japanese split
about 5 million years ago, human–chimpanzees split.

Phylogenetic tree for humans and closely related apes

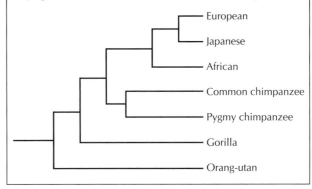

Ⓗ Cladistics

CLADES, CLADOGRAMS AND CLADISTICS

The tree diagrams shown on the previous page started to be produced in the second half of the 20th century. Neither the data on base or amino acid sequences nor the powerful computers needed to analyse the data were available before then. The diagrams use branching points, or **nodes**, to show groups of organisms which are related, and therefore presumably had common ancestry. These groups are called **clades**, from the Greek word klados – a branch.

A clade is a group of organisms that evolved from a common ancestor.

Clades can be large groups, with a common ancestor far back in evolution, or smaller groups with a more recent common ancestor.

The tree diagrams showing clades are called **cladograms**. Cladograms have been used to re-evaluate the classification of many groups of organisms. The methods used are very different from procedures traditionally used by taxonomists, so a new name has been given to this type of classification – **cladistics**.

Cladistics is a method of classification of living organisms based on the construction and analysis of cladograms.

CLADOGRAMS AND CLASSIFICATION

The classification of many groups has been re-examined using cladograms. In many cases, cladograms have confirmed existing classifications. This is not surprising as both traditional classification and cladistics are attempting to reflect phylogenetic relationships – the evolutionary origins of groups of living organisms.

Cladograms can be difficult to reconcile with traditional classifications, because the nodes can occur at any point. It can therefore seem rather arbitrary how the hierarchy of taxa is fitted to the clades.

In some cases, cladistics suggests radically different phylogenies. Should the existing classifications be trusted in these cases, or the new ones based on cladistics?

The strength of cladistics is that the comparisons between organisms are objective, based, as they are, on molecular differences. The weakness is that these molecular differences are analysed on the basis of probabilities. Occasionally improbable events occur, making the analyses wrong. So, although cladistics should not be treated as infallible, in many cases it can stimulate a reinterpretation of the data on which traditional classifications have been based.

CONSTRUCTING CLADOGRAMS

The construction of cladograms usually involves extremely complicated calculations that are done by powerful computers. The aim is to work out how the differences in base or amino acid sequence could have evolved with the smallest number of mutations. This is called parsimony analysis and although it does not prove how evolution did occur, it gives the most likely course. A simpler method of constructing a cladogram is given here. The amino acid sequence of hemoglobin has been compared in many vertebrates. The table (below) shows the numbers of differences in the amino acid sequence of ten vertebrates. The data in this table, and the details of what the amino acid differences are, has been used to construct the cladogram below. A time scale has been included by calibrating the rate of change in the amino acid sequences. By comparing the table and the cladogram, it is possible to deduce how a cladogram can be constructed from numbers of differences in base or amino acid sequence.

A simple cladogram could also be constructed using information about the form (morphology) of organisms.

Numbers of differences in the amino acid sequence of hemoglobin in ten vertebrates

	Elephant	Platypus	Ostrich	Starling	Crocodile	Lungfish	Coelacanth	Goldfish	Shark
Human →	26	40	43	41	47	83	70	68	71
Elephant →		45	45	48	50	84	72	63	74
Platypus →			54	52	51	89	74	70	76
Ostrich →				26	36	91	75	68	73
Starling →					47	91	77	67	70
Crocodile →						85	78	70	77
Lungfish →							90	94	86
Coelacanth →								83	78
Goldfish →									88

Phylogenetic tree diagram for ten vertebrates

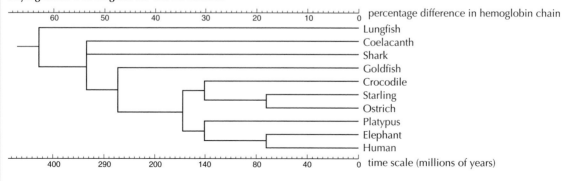

EXAM QUESTIONS ON OPTION D – EVOLUTION

D1 The scattergram below shows the relationship between brain size and total body mass in species of mammal. Primate species are shown as solid circles and other species of mammal as open circles.

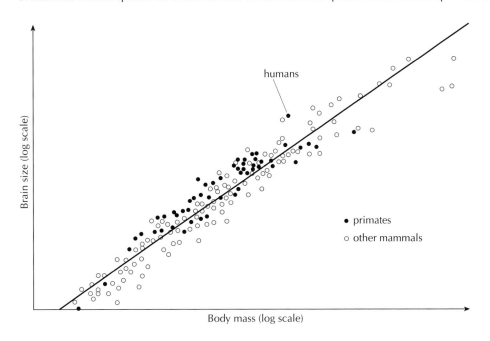

[Source: CUP, Encyclopaedia of Human Evolution,]

a) Using the data in the scattergram,

   (i) state the relationship between body mass and brain size in mammals   [1]

   (ii) compare the brain size in relation to body mass of primates with that of other mammals   [2]

   (iii) explain briefly how the scattergram can be interpreted to show that human brains are larger than those of other primates.   [2]

b) Increases in brain size in relation to body mass could be due either to increases in brain size or decreases in body mass. Suggest one advantage to primates of reduced body mass.   [1]

D2 The figure below shows the base sequence of part of a hemoglobin gene in four species of mammal.

Human	TGA CAA GAA CA - GTT AGA G - TGT CCGA
Orang utan	TCA CGA GAA CA - GTT AGA G - T GTC CGA
Lemur	TAA CGA TAA CAG GAT AGA G - T ATC TGA
Rabbit	TGG TGA TAA CAA GAC AGA GAT ATC CGA

a) Calculate the number of differences between base sequence of

   (i) humans and orang utans

   (ii) humans and lemurs

   (iii) humans and rabbits

   (iv) orang utans and lemurs

   (v) orang utans and rabbits

   (vi) lemurs and rabbits   [6]

b) Using the differences in base sequence between the four mammal species, construct a cladogram.   [4]

D3 In Africa, south of the Sahara and north of the Zambezi, the sickle cell allele Hb_s is very common. In some ethnic groups the proportion of newborn babies that are homozygous recessive can be as high as 0.053 (5.3%). These babies suffer from sickle cell anemia.

a) Calculate the **frequency** of the sickle cell allele in these ethnic groups.   [2]

b) Calculate the **percentage** of the population that are carriers of the sickle cell allele.   [1]

c) Outline the reasons for the high frequency of the sickle cell allele in these ethnic groups, despite the serious consequences of sickle cell anemia.   [2]

Stimulus and response

REFLEXES

One of the basic activities of the nervous system is the coordination of rapid responses to stimuli, including reflexes.

A stimulus is a change in the environment, either internal or external, that is detected by a receptor and elicits a response.

A response is a change in an organism, produced by a stimulus.

A reflex is a rapid unconscious response to a stimulus.

Although they are the simplest type of coordination, reflexes involve a precise pathway of neurons, with at least three synapses. The pathway is called a **reflex arc**. An example of a reflex is pulling away the hand after touching a hot object – this is called the pain withdrawal reflex. The reflex arc that coordinates this is shown in the diagram (below). Reflex arcs involve these five parts:

- **receptors** – to detect a stimulus; receptors can be sensory cells or nerve endings of sensory neurons
- **sensory neurons** – to receive messages across synapses, from receptors and carry them to the central nervous system (spinal cord or brain)
- **relay neurons** – to receive messages, across synapses, from sensory neurons, and pass them to the motor neurons that can cause an appropriate response
- **motor neurons** – to receive messages, across synapses, from relay neurons and carry them to an effector
- **effectors** – to carry out a response after receiving a message from a motor neuron; effectors can be muscles, which respond by contracting, or glands, which respond by secreting.

NATURAL SELECTION AND RESPONSES

Animal responses can be altered by natural selection if they are genetically programmed and affect the animal's chances of survival and reproduction. Offspring inherit successful types of response from their parents. Sometimes the environment of an animal species changes and natural selection may then favour a different type of response. Two examples related to global warming are given here, but there are many others from all around the world.

1. Migration in *Sylvia atricapilla* (blackcap)

This bird breeds in the early summer across much of central and northern Europe. It then migrates to warmer areas before the winter.

Until recently, populations in Germany migrated to Spain or other Mediterranean areas. Recent studies have shown a change in migration pattern, with 10% of the birds migrating to the UK.

Experiments with eggs have shown that the direction of migration is genetically programmed and inherited. The blackcaps that migrate from Germany to the UK for the winter instinctively tend to fly west, whereas those still migrating to Spain tend to fly southwest.

2. Timing of breeding in *Parus major* (great tit)

Parus major breeds in spring or early summer throughout much of Europe.

The timing of egg laying is genetically determined. Day length is used to determine the time of year.

Recent studies in the Netherlands have shown that the mean date of egg laying is becoming earlier. Adults that breed earlier enjoy greater reproductive success. This is due to the earlier opening of leaves on deciduous trees and an earlier peak in the biomass of invertebrates feeding on tree leaves. These invertebrates are the main food that adults collect and feed to offspring.

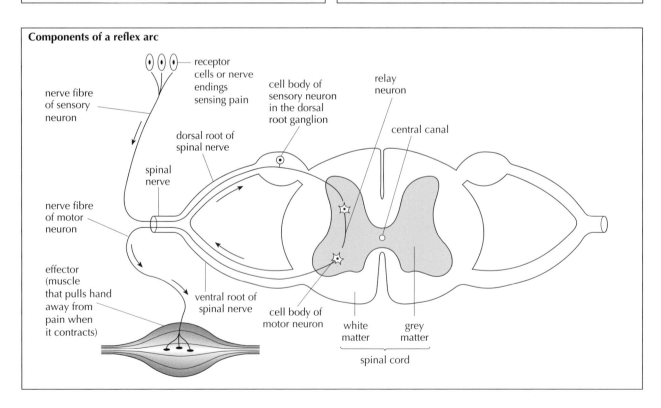

Components of a reflex arc

receptor cells or nerve endings sensing pain

nerve fibre of sensory neuron

cell body of sensory neuron in the dorsal root ganglion

relay neuron

central canal

dorsal root of spinal nerve

spinal nerve

nerve fibre of motor neuron

effector (muscle that pulls hand away from pain when it contracts)

ventral root of spinal nerve

cell body of motor neuron

white matter

grey matter

spinal cord

Perception of stimuli

DIVERSITY OF SENSORY RECEPTORS

Humans have a diversity of types of receptor and so can perceive a wide range of stimuli.

Type	Stimulus	Example
Mechanoreceptors	Mechanical energy in the form of sound waves Movements due to pressure or gravity	Hair cells in the cochlea of the ear Pressure receptor cells in the skin
Chemoreceptors	Chemical substances dissolved in water (tongue) Chemical substances as vapours in the air (nose)	Receptor cells in the tongue Nerve endings in the nose
Thermoreceptors	Temperature	Nerve endings in skin detect warm or cold
Photoreceptors	Electromagnetic radiation, usually in the form of light	Rod and cone cells in the eye

Structure of the human ear

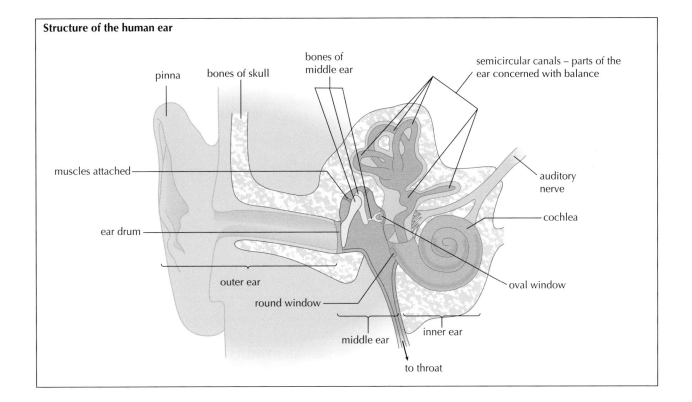

PERCEPTION OF SOUND

1. Eardrum

When sound waves reach the eardrum at the end of the outer ear, they make it vibrate. The vibration consists of rapid movements of the eardrum, towards and away from the middle ear. The role of the eardrum is to pick up sound vibrations from the air and transmit them to the middle ear.

2. Bones of the middle ear

There is a series of very small bones in the middle ear, called ossicles. Each ossicle touches the next one. The first ossicle is attached to the eardrum and the third one is attached to the oval window. The ossicles' role is to transmit sound waves from the eardrum to the oval window. They also act as levers, reducing the amplitude of the waves, but increasing their force, which amplifies sounds by about 20 times. The oval window's small size, compared with the eardrum, helps with amplification. Muscles attached to the ossicles protect the ear from loud sounds, by contracting to damp down vibrations in the ossicles.

3. Oval window

This is a membranous structure, like the eardrum. It transmits sound waves to the fluid filling the cochlea. This fluid is incompressible, so a second membranous window is needed, called the round window. When the oval window moves towards the cochlea, the round window moves away from it, so the fluid in the cochlea can vibrate freely, with its volume remaining constant.

4. Hair cells in the cochlea

The cochlea consists of a tube, wound to form a spiral shape. Within the tube are membranes, with receptors called hair cells attached. These cells have hair bundles, which stretch from one of the membranes to another. When the sound waves pass through the fluid in the cochlea, the hair bundles vibrate. Because of gradual variations in the width and thickness of the membranes, different frequencies of sound can be distinguished, because each hair bundle only resonates with particular frequencies. When the hair bundles vibrate, the hair cells send messages across synapses and on to the brain via the auditory nerve.

Vision in humans

PHOTORECEPTORS

The photoreceptors of the eye are contained in the retina. The figures (right) show the structure of the eye and the structure of the retina.

There are two types of photoreceptor cell – rod cells and cone cells. These cell types both absorb light and then transmit messages to the brain, via the optic nerve. They are different in these ways:

1. Rod cells are more sensitive to light than cone cells, so they function better in dim light. Rod cells become bleached in bright light, but cone cells function well.

2. Rod cells absorb all wavelengths of visible light, so they give monochrome vision, whereas there are three types of cone cell, sensitive to red, green and blue light, which give colour vision.

3. Groups of up to two hundred rod cells pass impulses to the same sensory neuron of the optic nerve, whereas many cone cells have their own individual neuron through which messages can be sent to the brain. Cone cells therefore give greater visual acuity than rod cells.

4. Rod cells are more widely dispersed through the retina so they give a wider field of vision, whereas cone cells are very concentrated near the fovea, giving one acute area of the field of vision.

Structure of the eye (in horizontal section)

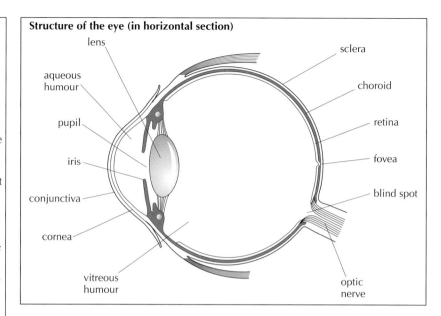

Structure of the retina

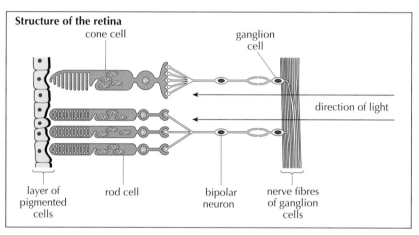

The Herman grid illusion

The grid (above) is a famous example of an optical illusion.

Grey areas appear at the intersections of the white lines, which are not real. If all of the grid is covered up apart from one white line, the grey areas disappear. This illusion can be explained in terms of the processing of visual stimuli.

PROCESSING OF VISUAL STIMULI

Between the perception of photons of light and impulses reaching the brain, there are a series of stages of processing of visual stimuli.

1. Convergence

Bipolar cells in the retina combine the impulses from groups of rod or cone cells and pass them on to ganglion cells (sensory neurons of the optic nerve).

2. Edge enhancement

Each ganglion cell is stimulated when light falls on a small circular area of retina called the receptive field. There are two types of ganglion cell. In one type, the ganglion is stimulated if light falls on the centre of the receptive field, but this stimulation is reduced if light also falls on the periphery. In the other type, light falling on the periphery of the receptive field stimulates the ganglion cell, but this stimulation is reduced if light also falls on the centre. Both types of ganglion cell are therefore more stimulated if the edge of light/dark areas is within the receptive field. White areas of the Herman grid look whiter if they are next to a black area.

Contralateral processing

The left and right optic nerves meet at a structure called the optic chiasma. Here all the neurons that are carrying impulses from the half of the retina nearest to the nose cross over to the opposite optic nerve. As a result the left optic nerve carries information from the right half of the field of vision and vice versa. This allows the brain to deduce distances and sizes.

Innate and learned behaviour

INNATE BEHAVIOUR IN INVERTEBRATES

Most behaviour in invertebrates is innate, not learned. Innate behaviour is sometimes called instinctive. It develops independently of the environmental context. In contrast, learned behaviour develops as a result of experience. Innate behaviour can be investigated by simple experiments with invertebrates, for example chemotaxis in *Planaria* (flatworms). A taxis is a movement towards or away from a directional stimulus. If *Planaria* are placed in a shallow dish with small pieces of food in part of the dish, they usually move towards the food. Other variables need to be kept constant, for example the amount of light in different parts of the dish. Also, in behaviour experiments like this, results should be quantitative, not merely descriptive. For example, a line could be drawn across the middle of the dish to mark the halves of the dish with and without food. The numbers of *Planaria* in each half of the dish could be recorded each minute during the experiment.

The graph (right) shows the results of an experiment, using slaters (woodlice), to investigate a behaviour pattern called kinesis. Kinesis is response to a non-directional stimulus, in which the rate of movement or the rate of turning depends on the level of the stimulus, but the direction of movement is not affected.

LEARNED BEHAVIOUR AND SURVIVAL

In diverse and changeable environments, animals can improve their chances of survival by learning new behaviour patterns. Examples:
- Some chimpanzees learn to catch termites by poking sticks into termite mounds.
- Birds learn to avoid eating orange and black striped cinnabar moth caterpillars, after associating their colouration and unpleasant taste.
- Many bird species learn to take avoiding action when they hear alarm calls warning them of a predator.
- Foxes learn to avoid touching electric fences after receiving an electric shock.
- In Britain, hedgehogs have learned to run across busy roads, instead of rolling up into a ball.

PAVLOV AND CONDITIONING IN DOGS

Ivan Pavlov investigated the salivation reflex in dogs. He observed that his dogs secreted saliva when they saw or tasted food. The sight or taste of meat is called the **unconditioned stimulus** and the secretion of saliva is called the **unconditioned response**.

Pavlov then gave the dogs a neutral stimulus, such as the sound of a ringing bell or ticking metronome, before he gave the unconditioned stimulus – the sight or taste of food. He found that after repeating this procedure for a few days, the dogs started to secrete saliva before they had received the unconditioned stimulus. The sound of the bell or the metronome is called the **conditioned stimulus** and the secretion of saliva before the unconditioned stimulus is the **conditioned response**.

The dogs had learned to associate two external stimuli – the sound of a bell or metronome and the arrival of food. This is called conditioning – an alteration in the behaviour of an animal as a result of the association of external stimuli.

Quantitative investigation of kinesis in slaters (woodlice)

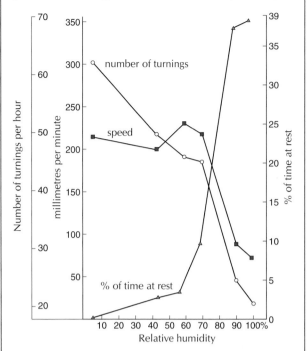

The graph shows that as humidity rises, the movement of the slaters is less and although the number of turns per hour is less, the number per metre moved is more. Slaters often congregate in small, humid spaces, increasing their chances of survival and reproduction.

DEVELOPMENT OF BIRDSONG

Birdsong is an interesting example of behaviour, because it has been shown in some species to be partly innate and partly learned. The chaffinch (*Fringilla coelebs*) is an example. Male chaffinches use their song to keep other males out of their territory and to attract females. The song varies a little between males, allowing identification of individuals. It also has recognizable features to show that it is a chaffinch singing. The figures (below) show the normal song of a male, reared where he could hear the song of adult chaffinches, and the song of male that was reared in isolation in a soundproof box. The song of the bird reared in isolation had some features of the normal song, including the correct length and number of notes, which must have been innate. However, there is a narrower range of frequencies, and fewer distinctive phases. These must be learned from other chaffinches.

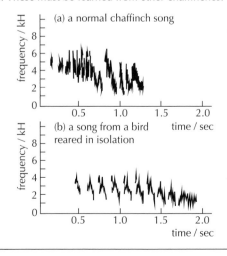

Neurotransmitters and synapses

EXCITATORY AND INHIBITORY SYNAPSES

Although there is a wide variety of synapses in the nervous system, especially the brain, there are two main types.

Excitatory synapses

This type of synapse was described in page 52. The neurotransmitter released by the presynaptic neuron causes sodium ions or other positively charged ions to enter the postsynaptic neuron, helping to depolarize it and cause an action potential. Postsynaptic transmission is therefore excited (stimulated).

Inhibitory synapses

In these synapses, the neurotransmitter released by the presynaptic neuron causes negatively charged chloride ions to move into the postsynaptic neuron, increasing its polarization. This effect, called hyperpolarization, makes it more difficult to depolarize a neuron sufficiently to cause an action potential. Postsynaptic transmission is therefore inhibited.

The electron micrograph (above right) shows adjacent neurons containing different neurotransmitters.

The graphs below show the effects of excitatory and inhibitory neurotransmitters on the membrane potential of a postsynaptic neuron.

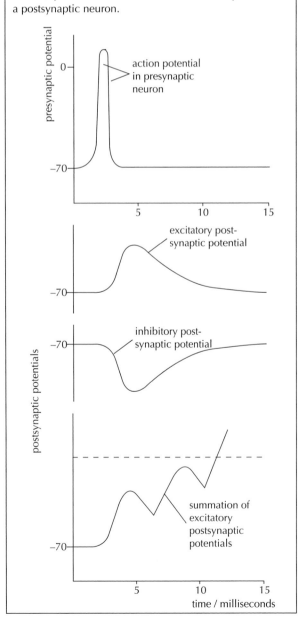

Electron micrograph showing adjacent neurons containing vesicles of different neurotransmitters

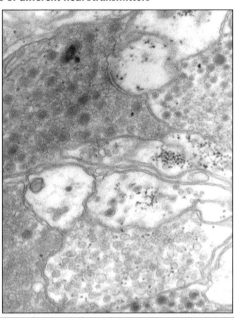

DECISION-MAKING IN THE CNS

One of the fundamental roles of the brain and spinal cord is decision-making. This can be a simple process, as in a reflex, or much more complicated, for example when choosing a partner. Synapses are the sites at which decisions are made.

One pulse of excitatory neurotransmitter, released when an action potential reaches the end of a postsynaptic neuron, is unlikely to be enough to cause postsynaptic transmission. A rapid sequence of pulses of neurotransmitter is needed. These could come from the same presynaptic neuron, or more likely from a number of different ones. This is possible because postsynaptic neurons have synapses with more than one pre-synaptic neurone, sometimes with hundreds.

Where many presynaptic neurons form synapses with a postsynaptic neuron, some of these synapses will be excitatory and others will be inhibitory. The effects of excitatory neurotransmitters may be cancelled out if an inhibitory neurotransmitter is also being released. Whether an action potential is initiated in the postsynaptic neuron is therefore decided by the summation of messages from all of these synapses. In this way, decisions can be made by the central nervous system. The figure (below) shows a postsynaptic motor neuron and some of its associated presynaptic neurons.

Synapses between many neurons and one motor neuron

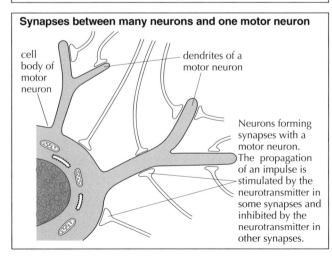

cell body of motor neuron

dendrites of a motor neuron

Neurons forming synapses with a motor neuron. The propagation of an impulse is stimulated by the neurotransmitter in some synapses and inhibited by the neurotransmitter in other synapses.

Psychoactive drugs

EXCITATORY AND INHIBITORY DRUGS

Drugs are chemical substances that are ingested, injected, inhaled or put into the body in some other way, to cause a change in the functioning of the body. Psychoactive drugs affect the brain and personality.

Most psychoactive drugs affect the functioning of the brain by disrupting synaptic transmission. Excitatory drugs work either by promoting transmission at excitatory synapses or inhibiting transmission at inhibitory synapses. Inhibitory drugs do the opposite.

Examples of psychoactive drugs

Excitatory	Inhibitory
nicotine	benzodiazepines
cocaine	alcohol
amphetamines	THC

Psychoactive drugs can affect synaptic transmission in a variety of ways:

- Some psychoactive drugs have a chemical structure similar to a neurotransmitter and so bind to receptors for that neurotransmitter in postsynaptic membranes. They block the receptors, preventing the neurotransmitter from having its usual effect.
- Other psychoactive drugs with a chemical structure similar to a neurotransmitter have the same effect as the neurotransmitter. However, unlike the neurotransmitter, they are not broken down so when they bind to the receptor the effect is much longer lasting.
- Some psychoactive drugs interfere with the breakdown of neurotransmitters in synapses or its reabsorption into the presynaptic neuron and so prolong the effect of neurotransmitters.

ADDICTION TO PSYCHOACTIVE DRUGS

The causes of addiction to psychoactive drugs have been widely studied, because of the physical and social damage that addictions can cause. Three factors increase levels of addiction, especially when they are combined.

1. Dopamine secretion

The first factor affecting whether addiction develops is the drug itself. Some drugs are addictive and some are not. A feature of many addictive drugs is that transmission is stimulated at synapses using dopamine as a neurotransmitter. These synapses are involved in the reward pathway, which gives us feelings of well-being and pleasure (see the example of cocaine below).

Users of addictive drugs find it very difficult to stop, because they have become dependent on the feelings that dopamine promotes.

2. Genetic predisposition

Even with many drugs that are potentially addictive, not everyone becomes an addict. Addictions, especially alcoholism, are much commoner in some families than others. This suggests that genes can make some people predisposed. Researchers are now trying to identify the genes that are involved.

3. Social factors

It is still not certain that a person who is genetically predisposed to develop addictions will do so when exposed to an addictive drug. Social factors can either prevent or encourage it.

Cultural traditions, peer pressure, poverty and social deprivation, traumatic life experiences and mental health problems all increase the chances of an addiction developing.

EFFECTS OF COCAINE AND THC

1. Cocaine

Cocaine is an excitatory psychoactive drug. It stimulates transmission at synapses in the brain that use dopamine as a neurotransmitter. Cocaine binds to membrane proteins that pump dopamine back into the presynaptic neuron. It blocks these transporters, causing a build-up of dopamine in the synapse.

The synapses that use dopamine as a neurotransmitter are responsible for pleasurable feelings that we get during certain activities, for example eating or having sex. Because cocaine causes continuous transmission at these synapses, it gives feelings of euphoria that are not related to any particular activity. It also causes users to have increased energy, alertness and talkativeness.

Cocaine is highly addictive and is a widely abused drug. Tissue taken from the brains of cocaine users after death had lower than normal levels of dopamine, suggesting that the body adapts to cocaine use by reducing secretion. This would explain cocaine-induced depression.

Crack is a form of cocaine that forms a vapour when it is heated. It can therefore be inhaled and absorbed very rapidly and gives very intense effects. These cause greater addiction and overdose problems than other forms of cocaine.

2. THC (tetrahydrocannabinol)

Cannabis contains a mixture of chemicals, but one of them called THC causes most of its psychoactive effects.

THC affects transmission at an unusual type of synapse, where the postsynaptic neuron can release a signalling chemical that binds to receptors in the membrane of the presynaptic neuron. It is not yet certain what these signalling chemicals are. THC also binds, giving them their name – cannabinoid receptors. When THC binds to cannabinoid receptors, it blocks the release of excitatory neurotransmitter. THC is therefore an inhibitory psychoactive drug.

Cannabinoid receptors are found in synapses in various parts of the brain, including the cerebellum, hippocampus and cerebral hemispheres.

Users make various claims about the effects of THC, most of which are not backed up by any evidence. There is good evidence for disruption of psychomotor behaviour so it is not safe to drive vehicles or operate machinery. Short-term memory impairment, intoxication and stimulation of appetite are other effects.

The human brain

FUNCTIONS OF PARTS OF THE BRAIN

The brain is made up of parts, each of which has a distinctive structure and carries out specific functions. The structure of the brain is shown in the diagram (right). The parts labelled in the diagram have these functions:

Medulla oblongata – controls automatic and homeostatic activities, such as swallowing and vomiting, digestion, breathing and heart activity

Cerebellum – coordinates unconscious functions such as balance and movements, including hand-eye coordination.

Hypothalamus – maintains homeostasis using both the nervous and endocrine systems; produces the hormones that are secreted by the posterior pituitary gland; sends releasing factors to stimulate hormone secretion by the anterior pituitary gland

Pituitary gland – posterior lobe stores and secretes hormones produced by the hypothalamus; anterior lobe produces and secretes hormones that regulate many body functions

Cerebral hemispheres – receives impulses from the eye, ear, nose and tongue; acts as the integrating centre for higher complex functions, including learning, memory, emotions and consciousness.

INVESTIGATING BRAIN FUNCTION

Various techniques have been used to find out what the function of each part of the brain is.

1. Animal experiments

Many experiments have been performed on animals, including primates, often involving surgical procedures – parts of the skull have to be removed to get access to the brain. The animal must be kept alive so that the brain is still functioning. Experimental procedures are carried out on the brain and the effects on the animal are then observed, either during the operation or afterwards.

Many scientists have ethical objections to these experiments as the animals may experience some suffering and are often sacrificed.

2. Lesions

Accidents, strokes and tumours can damage specific parts of the brain. The damaged areas are called lesions and from them, the location of particular brain functions can be deduced. For example, lesions in Broca's area in the left cerebral hemisphere cause dysphasia – inability to speak, but reading and writing are still possible. The craving-centre of the brain was first identified from the case of a man who lost the desire to smoke cigarettes, after a stroke damaged a region in his brain called the insula.

3. Functional magnetic resonance imaging

Functional magnetic resonance imaging (fMRI) is a technique for determining which parts of the brain are activated by specific thought processes. Active parts of the brain receive increased blood flow, which fMRI records.

The experimental subject is placed in the scanner and a high-resolution scan of the brain is taken. A series of low-resolution scans is then taken, while the subject is being given a stimulus. The scans show which parts of the brain are activated during the response to the stimulus. An example of fMRI is shown (right). It indicates activity in the visual cortex.

Structure and function of parts of the brain

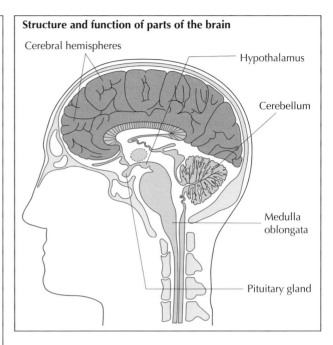

THE PUPIL REFLEX AND BRAIN DEATH

If a bright light shines into one eye, the pupils of both eyes constrict. This is called the pupil reflex. Photoreceptor cells in the retina detect the light stimulus. Nerve impulses are sent in sensory neurons of the optic nerve to the brain. The medulla oblongata (brain stem) processes the impulses and then sends impulses to circular muscle fibres in the iris of the eye. These muscle fibres contract, causing the pupil to constrict.

In the past, when a vital organ of the body ceased to function, the whole body would rapidly die. Advances in medicine now allow the rest of the body to be kept alive when certain organs are not functioning. Sometimes an organ of the body recovers after a time and the patient can enjoy a good quality of life again.

If a patient is in a coma (prolonged unconsciousness) because of damage to the cerebral hemispheres, recovery may be possible. However, damage to the medulla oblongata is much more serious and recovery cannot be expected. Doctors therefore use tests of brain stem function to decide whether to try to preserve a patient's life. The pupil reflex is often used. If an unconscious patient's pupils do not constrict when a light is shone into the eye, this suggests that they have injuries serious enough to have caused brain death.

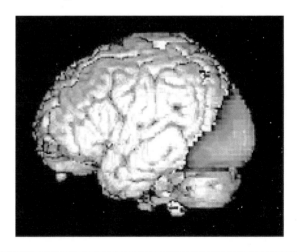

UNCONSCIOUS COORDINATION

The part of the nervous system that is used to control internal processes unconsciously is called the autonomic nervous system. Impulses are sent from the brain through the two parts of this system – the parasympathetic and sympathetic systems. Three of the processes controlled by these two systems are outlined in the table (below).

Organ	Parasympathetic system	Sympathetic system
Heart	Heart rate is slowed as the body is relaxed and less blood flow is needed.	Heart rate speeds up so that more blood can be pumped to the muscles.
Blood flow to the gut	Blood vessels are dilated, increasing blood flow to the gut.	Blood vessels are constricted, decreasing blood flow to the gut.
Iris of the eye	Circular muscle fibres contract, so the pupil constricts to protect the retina.	Radial muscles contract, dilating the pupil to give a better image.

PERCEPTION OF PAIN

Pain receptors are located in the skin and other organs. They consist of free nerve endings, which perceive mechanical, thermal or chemical stimuli. Impulses are sent from these pain receptors to sensory areas of the cerebral cortex, causing feelings of pain.

These feelings are necessary to allow us to know when our body is being damaged, so that we can take avoiding action – pain withdrawal reflexes for example.

However, pain sometimes becomes excessive or stops us from concentrating on important activities. In these situations, the pituitary gland releases **endorphins**. The endorphins are carried in the blood to the brain. They bind to receptors in the membranes of neurons that send pain signals and block the release of a neurotransmitter that is used to transmit the pain signals within the brain.

Endorphins are secreted during stressful times, after injuries and even during physical exercise such as running.

Male red deer fighting (rutting) in the fall

FORAGING BEHAVIOUR

When animals search for food, they are foraging. Research has shown that animals optimize food intake by their foraging behaviour. Two examples of this are described here.

1. **Bluegill sunfish** (*Lepomis macrochirus*)
These fish live in ponds, where they prey on small invertebrates, including *Daphnia*. When there is a low density of prey, bluegill sunfish consume all sizes of them. At medium prey densities, bluegill sunfish consume only prey of moderate or larger sizes. At high prey densities they mostly consume large prey, plus some of medium size. Consuming small numbers of large prey takes less energy than large numbers of small prey, hence the preference for large prey. At low prey densities, smaller prey have to be eaten as well, to get enough food in total.

2. **Starlings** (*Sturnus vulgaris*)
Starlings are birds that feed their young mainly on crane-fly larvae, which they obtain by probing into soil with their beak. Starlings become less efficient at probing for larvae, as the number of larvae they are holding in their beaks increases. The fewer journeys back to the nest, the less time and energy is used in transporting the larvae to the offspring.

The optimum number of larvae for starlings to catch and carry back to the nest depends on the distance between the foraging area and the nest. As the distance increases, the optimum number of larvae also increases.

When starlings have been observed, the number of larvae actually caught and transported has been found to be very close to the theoretical optimum.

RHYTHMICAL BEHAVIOUR PATTERNS

Many animals show rhythmical patterns in behaviour. These usually follow either a diurnal (daily) or an annual (yearly) cycle. One example of each is given here. There are longer cycles, for example the 13 and 17-year reproductive cycles of cicadas.

1. **Moonrats**
Like many mammals, moonrats (*Echinosorex gymnura*) are nocturnal. They live in Asia, in lowland forests including mangroves.

Their excellent sense of smell helps them to forage at night when much of their prey is active – insects and other invertebrates. They are less vulnerable to predation at night and in the day they rest in holes among tree roots or in hollow logs, where they are unlikely to be discovered.

2. **Red deer**
Reproduction follows an annual cycle in red deer (*Cervus elaphus*). Males and females are only sexually active in the fall (autumn).

Males fight to establish dominance over groups of females (figure, left) with whom they mate. The advantage is that if the females start gestation in the fall, the offspring are born in spring. Most food is available in spring and summer for feeding the offspring, so this type of season breeding gives the offspring the greatest chance of survival.

In the fall, males try to take possession of as large a group of females as they can and mate with them when they come into oestrus.

SOCIAL ORGANIZATION

Some animals live in colonies with clear social organization. Two examples are described here:

1. **Honey bees** live in colonies consisting of up to sixty thousand individuals. The colony acts like a super-organism that lives or dies together, and can reproduce to form extra colonies by swarming. There are three castes of honey bee, each of which has different tasks. The single queen bee is normally the only member of the colony to lay eggs. The worker bees do all the jobs that are needed to maintain the colony. The drones do nothing to help the colony to survive, but if they successfully mate with virgin queens they spread the genes of the colony to new colonies. Workers eject drones from the colony at the end of the season in which virgin queens are available. The table below summarizes the tasks of the three castes.

Caste	Gender	Tasks
Queen	Fertile female	Laying eggs. Producing a pheromone to control the activities of workers.
Drone	Fertile male	Mating with virgin females.
Worker	Infertile female	Collecting nectar and pollen. Converting pollen into honey. Secreting wax and using it to build the comb. Feeding and looking after larvae. Guarding the hive.

Waggle dance of honey bees

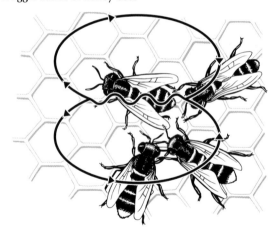

2. **Naked mole rats** (*Heterocephalus glaber*) live in colonies of up to 80 individuals, in burrow systems in parts of East Africa. One dominant female mole rat acts like a queen bee. She is the only female in the community to reproduce, mating with one of the males in the colony. Three other castes of mole rat help her.
   • 'Frequent workers' dig the tunnels and bring food.
   • 'Infrequent workers' are larger and occasionally help with heavier tasks.
   • 'Non-workers' live in the central nest, keeping the breeding female and her young offspring warm and defending the colony if it is attacked.
   The large and complex burrow systems could probably not be constructed or defended without social organization.
   A colony of social organisms is sometimes considered to be one super-organism. Either the colony as a whole survives and reproduces to form new colonies or it does not. Natural selection therefore exists at the level of the colony.

EVOLUTION OF ALTRUISTIC BEHAVIOUR

A dictionary definition of altruism is simply 'unselfish behaviour'. In Biology it has come to mean something more specific – altruism is defined as actions that increase another individual's lifetime number of offspring at a cost to one's own survival and reproduction. Parental care is therefore not altruism. There has been much discussion about the evolution of altruistic behaviour. We might expect natural selection always to be against behaviour that reduces the chances of survival and reproduction, yet there are some well-known examples of altruistic behaviour.

1. **Non-breeding naked mole rats**
The tasks of non-breeding workers in a naked mole rat colony are described (left). These tasks allow the breeding male and female in the colony to reproduce successfully. The evolution of this type of altruism, sometimes called kin selection, is easy to explain. The mole rats in a colony are all genetically related, so although the workers are helping to rear offspring that are not their own, they are helping to ensure the survival of their own genes.

2. **Blood sharing in vampire bats**
This behaviour was investigated in a population of vampire bats in Costa Rica. They live in groups and feed at night by sucking blood from larger animals. If one of the bats in the group fails to feed for more than two consecutive nights it may die of starvation. However, bats that have fed successfully regurgitate blood for a bat that has failed to feed. Tests have shown that this is done whether the two bats are genetically related or not. This is called reciprocal altruism because the bat that donates food to a hungry bat may in the future receive blood when it is hungry. There is an advantage for the whole group, because the benefit of receiving blood when starving is greater than the cost of donating blood after feeding well.

EVOLUTION OF EXAGERRATED TRAITS

Some species of animal have characteristics or behaviour patterns that seem to be developed excessively. The long and brightly coloured tail feathers of a peacock are an example. These are only used during courtship, to try to attract a female. At other times, the tail feathers will be an encumbrance, hindering rapid movement, especially during attacks by predators. This may be the explanation for the evolution of an exaggerated trait: any individual that survives, despite the exaggerated trait, must be well-adapted in other ways and so is a good mate to choose.

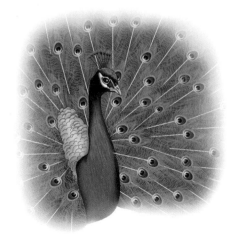

E1 Odorants are substances which can be detected by chemoreceptors in the nose. Many different odorants can be detected but each chemoreceptor cell is sensitive to only one type. The diagrams (right) show the mechanism used in the chemoreceptor.

a) Deduce which part of the mechanism is different in chemoreceptor cells that are sensitive to different odorants. [2]

b) When the odorant binds to the receptor protein, the receptor protein starts activating G protein. Using the data shown in the diagrams outline the effects of activated G protein. [3]

c) Predict the effect of entry of calcium ions and exit of chloride ions on the chemoreceptor cell. [1]

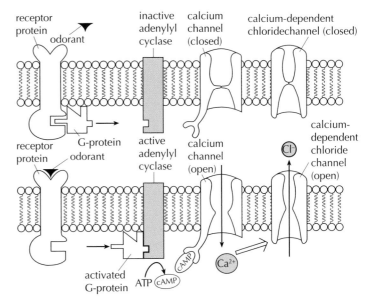

[Source of data: Gold et al, *Nature*, (1997), 385, page 677]

E2 a) State which type of receptor is found in the eye. [1]

b) Outline the neural pathway involved in the pupil reflex. [2]

c) State how this reflex can be used to find out the condition of the central nervous system. [1]

E3 The electron micrographs (centre and right) show the structure of hair cells in the cochlea.
The scanning electron micrograph (centre) shows two cells, each sitting in the cup-shaped upper surface of a Dieter's cell.

a) Suggest a function for the strut-like projection from the Dieter's cells. [1]

b) Perception of sound depends on movement of the hairs (stereocilia) that project from the upper surface of the hair cells.

(i) Describe the group of stereocilia projecting from one hair cell. [3]

(ii) Calculate the length of the longest stereocilium. The scale bar is 5 µm long. [1]

(iii) Explain how the stereocilia of hair cells that perceive high and low frequency sounds would differ. [1]

c) There are two types of hair cell in the cochlea, inner and outer. The hair cells shown in the micrographs are outer hair cells. They do not pass impulses directly to neurones, but act like cellular pistons, lengthening and shortening at the same frequency as the sound that they perceive.

(i) Suggest how the action of outer hair cells helps in hearing. [2]

(ii) Deduce, from the structure of the hair cell shown in the transmission electron micrograph (right), where energy is expended to shorten and lengthen the cell. [2]

Classification of microbes

CLASSIFICATION IN THREE DOMAINS

A system of classification of living organisms into five kingdoms was developed in the second half of the 20th century. Biologists mostly accepted it. In this classification, all prokaryotes were placed in one kingdom and eukaryotes in four kingdoms. However, when the base sequence of nucleic acids was compared, two very different groups of prokaryotes were identified. These groups are as different from each other as from eukaryotes. A higher grade of taxonomic group was needed to reflect this, now called a **domain**. Three domains have been described:
• Archaea
• Eubacteria
• Eukaryota.

The original evidence for this came from base sequences of ribosomal RNA, which is found in all organisms and evolves slowly, so it is suitable for studying the earliest evolutionary events. More evidence has since been obtained from gene sequencing studies. The table (right) shows the fundamental differences that also justify the new classification. An explanation of introns is given in page 61. The figure (below) shows the classification of living organisms into three domains. Names of groups within each domain have been omitted.

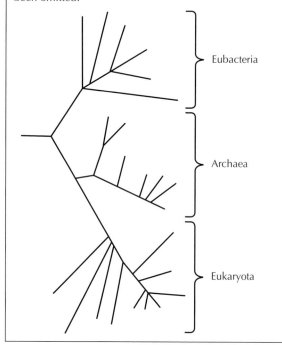

DISTINGUISHING THE THREE DOMAINS

Significant characters that are useful in distinguishing between the three domains	Archaea	Eubacteria	Eukaryota
Are cell walls made of peptidoglycan?	None	All	None
What are the bonds in membrane lipids?	Ether	Ester	Ester
What size are ribosomes?	70S	70S	80S
Do most genes contain introns?	No	No	Yes
How many species have histone proteins?	A few	None	All

CELL WALLS IN EUBACTERIA

Cell wall structure varies in the Eubacteria. There are two main types of structure, which are shown in the diagrams (below). Eubacteria with the structure shown in the upper diagram are called Gram-positive, as they are stained purple by Gram stain, whereas Eubacteria with the structure shown in the lower diagram are Gram-negative, as they stain less intensively and appear red.

Gram-positive Eubacteria

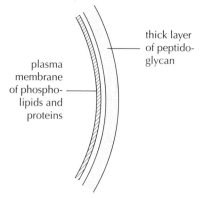

thick layer of peptido-glycan

plasma membrane of phospho-lipids and proteins

Gram-negative Eubacteria

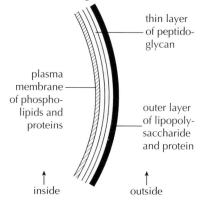

thin layer of peptido-glycan

plasma membrane of phospho-lipids and proteins

outer layer of lipopoly-saccharide and protein

inside   outside

HABITATS OF ARCHAEA

The Archaea are very diverse in their metabolism and this helps them to thrive in a very wide diversity of habitats, including some of the more extreme in the world.
• **Halophiles** live in habitats with a very high salt content – at least 1.5 mol dm^{-3} and often much higher. These concentrations are found in saline lakes such as the Dead Sea.
• **Thermophiles** live in very hot habitats, up to 100°C in some cases. Examples of these habitats are hot springs in volcanic areas and geothermally heated regions of the sea floor, including hydrothermal vents known as black smokers.
• **Methanogens** live in anaerobic habitats where organic matter is available. Examples are swamps and waterlogged soils, in the gut of cattle and other ruminants and in dumps of organic waste created by humans.

Diversity of microbes

SHAPE IN EUBACTERIA

The cells of Eubacteria vary considerably in shape. They can be spherical, rod-shaped, spiral, or comma shaped, for example.

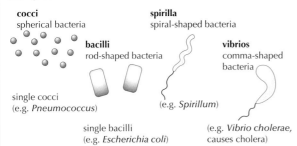

cocci
spherical bacteria

single cocci
(e.g. *Pneumococcus*)

bacilli
rod-shaped bacteria

single bacilli
(e.g. *Escherichia coli*)

spirilla
spiral-shaped bacteria

(e.g. *Spirillum*)

vibrios
comma-shaped bacteria

(e.g. *Vibrio cholerae*, causes cholera)

Although bacteria can exist as single cells, some species can also form aggregates – groups of cells linked together. For example, layers of bacteria called biofilms can form on rocks or other surfaces. The cells jointly secrete adhesive polysaccharides, sticking the cells to the surface and to each other. Single cells could not produce enough of the polysaccharide for efficient adhesion. The bacterium *Streptococcus mutans* forms biofilms on teeth, which are called plaque and can cause dental decay.

STRUCTURE OF VIRUSES

Most biologists do not consider viruses to be living organisms. Instead, they are regarded as genetic structures that can reproduce using the cells of a living organism. Every virus has a small number of genes composed of nucleic acid, surrounded by a protein coat. The coat is called the capsid. Apart from this, there are few similarities in structure. Viruses probably are diverse in structure because they evolved repeatedly, rather than all evolving from a single ancestral virus.

There are three key differences in virus structure:

1. Is the capsid enveloped?
In many viruses, the capsid is naked – it is the outer layer. In other viruses, there is a lipid bilayer outside the capsid. These are called enveloped viruses.

2. Are the genes DNA or RNA?
The genes in some viruses are composed of DNA whereas in others they are RNA.

3. Are the genes single or double-stranded?
The genes of viruses can be either single stranded or double stranded, whether they are composed of DNA or RNA.

DIVERSITY OF MICROSCOPIC EUKARYOTES

Any living organism that is too small to see with the naked eye is microscopic. There are many types of microscopic eukaryote, which are very diverse in their modes of nutrition, their modes of locomotion and whether they have cell walls, chloroplasts and cilia or flagella. Five types are described below.

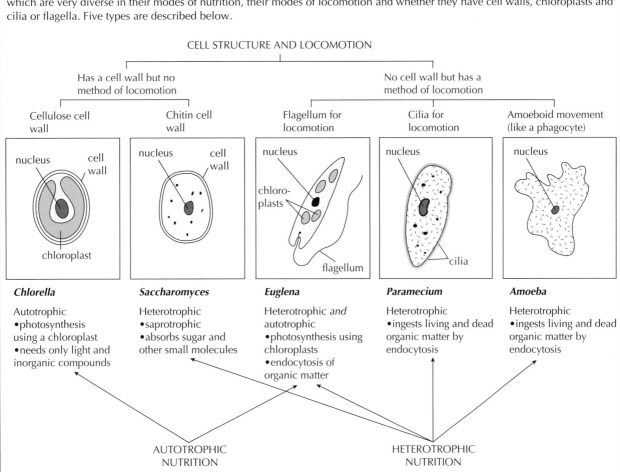

CELL STRUCTURE AND LOCOMOTION

Has a cell wall but no method of locomotion

No cell wall but has a method of locomotion

| Cellulose cell wall | Chitin cell wall | Flagellum for locomotion | Cilia for locomotion | Amoeboid movement (like a phagocyte) |

Chlorella

Autotrophic
•photosynthesis using a chloroplast
•needs only light and inorganic compounds

Saccharomyces

Heterotrophic
•saprotrophic
•absorbs sugar and other small molecules

Euglena

Heterotrophic *and* autotrophic
•photosynthesis using chloroplasts
•endocytosis of organic matter

Paramecium

Heterotrophic
•ingests living and dead organic matter by endocytosis

Amoeba

Heterotrophic
•ingests living and dead organic matter by endocytosis

AUTOTROPHIC NUTRITION

HETEROTROPHIC NUTRITION

The nitrogen cycle

ECOLOGICAL ROLES OF MICROBES

Microbes are very varied in their metabolism and so can have many different roles in ecosystems. For example, some microbes are **producers**. Other microbes are **decomposers**. Microbes can also be **nitrogen fixers** in ecosystems.

MICROBES AND THE NITROGEN CYCLE

Many microbes, including nitrogen fixers, have roles in the nitrogen cycle. The whole cycle is shown in the diagram (below).

1. Nitrogen fixation

Free-living *Azotobacter* and *Rhizobium* living mutualistically in root nodules both fix nitrogen. Nitrogen fixation is conversion of nitrogen from the atmosphere into ammonia, using energy from ATP.

2. Nitrification

The conversion of ammonia to nitrate (nitrification) involves two types of soil bacteria. *Nitrosomonas* converts ammonia to nitrite and *Nitrobacter* convert nitrite to nitrate.
Nitrification happens very rapidly, as long as soils are well aerated with abundant supplies of oxygen.

3. Denitrification

Nitrate is sometimes converted into nitrogen in a type of anaerobic respiration. This process is called denitrification as it reduces nitrate levels in soils.
Pseudomonas denitrificans is an example of a bacterium that carries out denitrification. Nitrate is broken down when it is used as a terminal electron acceptor in respiration instead of oxygen.
Anaerobic soils therefore encourage denitrification. Bad drainage and waterlogging are a frequent cause of anaerobic conditions in soils.

NITRATE FERTILIZERS AND RIVERS

Although nitrate is an essential nutrient for plants, its ecological effects in rivers can be detrimental.

1. Nitrate ions are soluble and are leached from soils very easily if excessive amounts are applied to crops. If phosphate and other minerals also reach a high concentration, a river becomes **eutrophic**.
2. The eutrophication causes algae to proliferate. Nitrate from fertilizers sometimes causes an excessive growth of algae, called an **algal bloom**. Some of the algae are deprived of light and die.
3. Bacteria decompose the dead algae. The bacteria create an increased **biochemical oxygen demand** and so cause **deoxygenation** of the water.
4. Low oxygen levels kill fish and other aquatic animals.

The graph (below) shows the results of an experiment in which a lake was fertilized with nitrates every year starting in 1969. The density of algae was estimated by measuring chlorophyll concentration in the water. Before the experiment, the concentration was below $5\,\mu g\,dm^{-3}$ throughout the year.

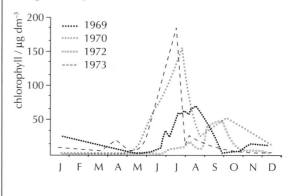

The nitrogen cycle

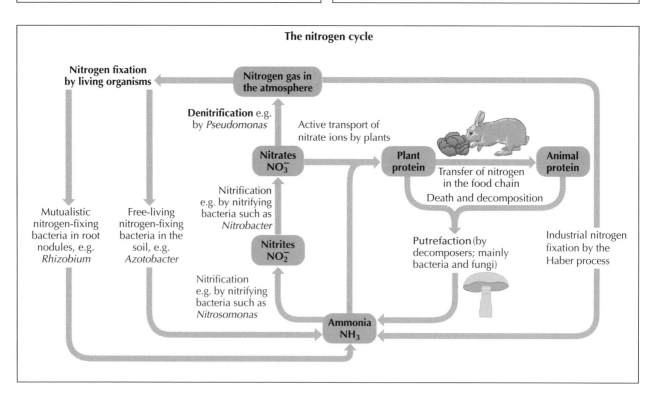

Sewage treatment and methane generation

SEWAGE IN RIVERS

Raw sewage often contains pathogens. If it is released into rivers, and people drink water from the river or swim in it, they may become infected with the pathogens.

Raw sewage also has ecological effects on rivers, shown in the figure (below).

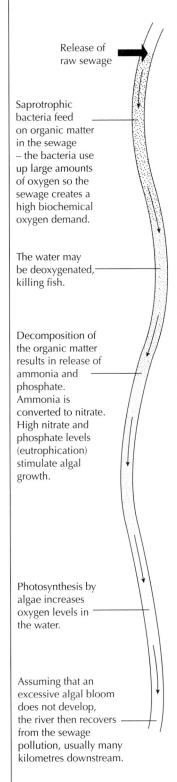

Release of raw sewage

Saprotrophic bacteria feed on organic matter in the sewage – the bacteria use up large amounts of oxygen so the sewage creates a high biochemical oxygen demand.

The water may be deoxygenated, killing fish.

Decomposition of the organic matter results in release of ammonia and phosphate. Ammonia is converted to nitrate. High nitrate and phosphate levels (eutrophication) stimulate algal growth.

Photosynthesis by algae increases oxygen levels in the water.

Assuming that an excessive algal bloom does not develop, the river then recovers from the sewage pollution, usually many kilometres downstream.

USING MICROBES IN SEWAGE TREATMENT

1. Trickle filter beds

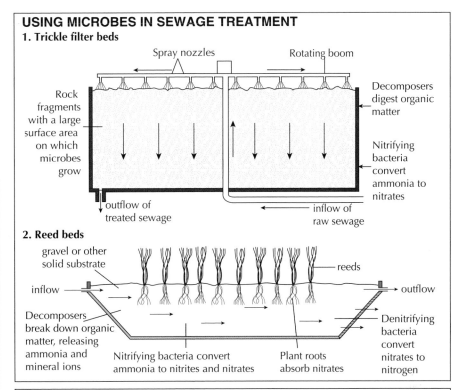

Spray nozzles

Rotating boom

Rock fragments with a large surface area on which microbes grow

Decomposers digest organic matter

Nitrifying bacteria convert ammonia to nitrates

outflow of treated sewage

inflow of raw sewage

2. Reed beds

gravel or other solid substrate

reeds

inflow

outflow

Decomposers break down organic matter, releasing ammonia and mineral ions

Nitrifying bacteria convert ammonia to nitrites and nitrates

Plant roots absorb nitrates

Denitrifying bacteria convert nitrates to nitrogen

METHANE GENERATION

Biomass already provides large amounts of fuel, in the form of wood, crop residues and dried manure. Methods now exist for converting biomass into fuels that are more convenient to use, such as ethanol and methane.

Methane is sometimes called marsh gas, because it is naturally produced by microbes in anaerobic conditions. These conditions are recreated in **bioreactors** used for methane generation.

A variety of types of organic matter can be the feedstock, including manure from farm animals and cellulose. The feedstock is loaded into the bioreactor where anaerobic conditions encourage the growth of three groups of naturally occurring bacteria.

The first group convert organic matter into organic acids and alcohol.

The second group convert organic acids and alcohol into carbon dioxide, hydrogen and acetate.

The third group of bacteria are the **methanogenic archaea** – they produce methane from carbon dioxide, hydrogen and acetate.

$$\text{Carbon dioxide} + \text{hydrogen} \rightarrow \text{methane} + \text{water}$$
$$CO_2 + 4H_2 \rightarrow CH_4 + 2H_2O$$

$$\text{Acetate} \rightarrow \text{methane} + \text{carbon dioxide}$$
$$CH_3COOH \rightarrow CH_4 + CO_2$$

The gas that is produced in bioreactors is sometimes called biogas and is 40–70% methane. It is renewable fuel. Production of it helps to dispose of potentially polluting organic wastes.

A bioreactor

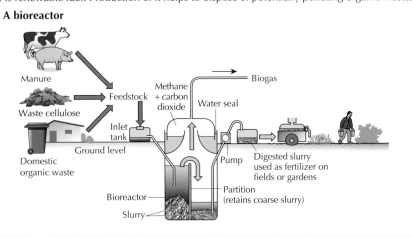

Manure

Waste cellulose

Domestic organic waste

Feedstock

Inlet tank

Ground level

Methane + carbon dioxide

Water seal

Biogas

Pump

Digested slurry used as fertilizer on fields or gardens

Partition (retains coarse slurry)

Bioreactor

Slurry

Microbes and biotechnology

GENE THERAPY

Gene therapy is the treatment of genetic disease by altering the genotype. In the case of a disease caused by a recessive allele, a fully functional dominant allele must be inserted into defective cells. The situation is more complex with genetic diseases caused by dominant alleles – expression of the defective gene must be prevented and a functioning allele may also need to be inserted.

There are two stages in the life cycle when gene therapy could be attempted:

1. **Somatic-cell therapy** – body cells are altered, which do not develop into gametes. Often very large numbers of cells will need to be altered, and although the genetic disease may be cured in the treated individual, it can still be passed on to offspring.

2. **Germ-line therapy** – sperm or egg cells are treated (or cells that will divide to produce sperm or egg cells). The disease should be completely absent in offspring formed using the gametes, but the parent in which the disease has been diagnosed still has the disease.

VIRAL VECTORS IN GENE THERAPY

Viruses have had millions of years to evolve efficient mechanisms for entering mammalian cells and delivering genes to them. They sometimes also incorporate these genes into the host cell's chromosomes. Viruses are therefore obvious candidates for the gene delivery system, needed in gene therapy. Modified viruses must be produced containing the desired gene, which will infect target cells but not replicate to form more virus particles. A modified virus that is used in this way is called a vector.

The most widely used virus vectors are retroviruses.
One example of their use is in the treatment of SCID (severe combined immuno-deficiency), a genetic disease that is due to the lack of an enzyme called ADA. A famous early case involved a baby called Andrew:

Gene therapy for SCID

Genetic screening before birth shows that Andrew has SCID

The allele that codes for ADA is obtained. This gene is inserted into a retrovirus

Blood removed from Andrew's placenta and umbilical cord immediately after birth contains stem cells. These are extracted from the blood

Retroviruses are mixed with the stem cells. They enter them and insert the gene into the stem cells' chromosomes

Stem cells containing the working ADA gene are injected into Andrew's blood system via a vein.

For four years T-cells (white blood cells), produced by the stem cells, made ADA enzymes, using the ADA gene. After four years more treatment was needed.

RISKS OF GENE THERAPY

Most attempts at gene therapy so far have not been successful and the hopes of patients and their families have been raised and then disappointed. There have also been cases where the treatment has harmed the patients. One example of this involved a trial of gene therapy for SCID using retroviruses, in a group of ten children in France. Two of the children developed leukemia. The viral vector had inserted DNA into a cancer-causing gene and activated it. Adenoviruses are possible alternative viral vectors, as they do not insert their genes into host cell chromosomes, so should not activate the cancer-causing genes.

USE OF REVERSE TRANSCRIPTASE IN MOLECULAR BIOLOGY

Retroviruses, such as HIV, are viruses that use RNA as their genetic material. They contain an enzyme that catalyses the production of DNA from RNA. This enzyme is called reverse transcriptase. Retroviruses use it to make a DNA copy of their RNA genes, after they have entered a host cell. The DNA copy becomes inserted into the host cell's chromosomes.

Molecular biologists use reverse transcriptase to make copies of the genes that they use in gene transfer.

1. Cells that are transcribing the required gene are obtained and mRNA transcripts are extracted.

2. Single-stranded DNA copies of the mRNA are made using reverse transcriptase. This is called cDNA.

3. DNA polymerase is used to convert the single-stranded DNA into double-stranded DNA, producing genes that can be transferred into another organism.

The figure (below) is a summary of the procedure. The genes produced contain no introns, so if they are transferred to bacteria, which do not edit out introns, the correct protein will nonetheless be produced.

Production of cDNA from mRNA

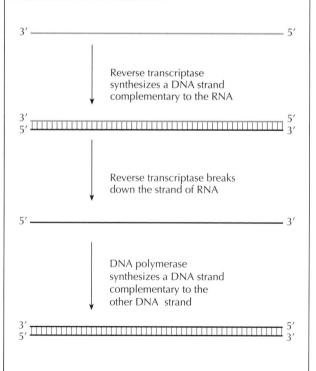

3′ ———————————————————— 5′

Reverse transcriptase synthesizes a DNA strand complementary to the RNA

3′ ▯▯▯▯▯▯▯▯▯▯▯▯▯▯▯▯▯▯▯ 5′
5′ ▯▯▯▯▯▯▯▯▯▯▯▯▯▯▯▯▯▯▯ 3′

Reverse transcriptase breaks down the strand of RNA

5′ ———————————————————— 3′

DNA polymerase synthesizes a DNA strand complementary to the other DNA strand

3′ ▯▯▯▯▯▯▯▯▯▯▯▯▯▯▯▯▯▯▯ 5′
5′ ▯▯▯▯▯▯▯▯▯▯▯▯▯▯▯▯▯▯▯ 3′

Microbes and food production

SACCHAROMYCES AND BREWING

Beer, wine and other alcoholic drinks are brewed using yeasts from the genus *Saccharomyces*. When oxygen is absent, the yeast switches to anaerobic cell respiration and excretes the ethanol and carbon dioxide produced in this process.

Anaerobic cell respiration is also called fermentation. During fermentation, the ethanol concentration of the fluid around the yeast cells can rise to approximately 15% by volume, before it becomes toxic to the yeast and the fermentation ends. Most of the carbon dioxide bubbles out into the atmosphere.

1. Wine production
Grapes are crushed to make juice, which is placed into large vessels, or vats. Yeast cells are naturally present on the grapes and they grow and divide in the juice. Sometimes selected varieties of yeast are added. Within a few days, all of the oxygen in the juice is used up and the yeast cells respire anaerobically from then onwards. The fermentation ends when the sugar in the juice has been used up or when the ethanol content reaches 15%. The juice has then become wine.

2. Beer production
Starch, rather than sugars, are the feedstock for beer production. Yeast cannot ferment starch, so there must be an extra stage in the production process. Barley seeds are wetted and allowed to start germinating. Amylase is produced in the germinating seeds. The amylase can convert starch to maltose. The barley seeds are dried after a few days to kill them and preserve the amylase. The dried, semi-germinated barley seeds are called malted barley.
Beer production involves mixing malted barley, with other sources of starch, and a selected variety of yeast into water in a vat. Hops are added as flavouring. Amylase from the malted barley digests starch, to release maltose, which the yeast converts to ethanol by anaerobic cell respiration.

SACCHAROMYCES AND BAKING

Yeast is used in baking bread. It is mixed into the dough before baking. The yeast produces ethanol and carbon dioxide by anaerobic cell respiration. The carbon dioxide forms bubbles within the dough, making the dough rise – it increases in volume. This makes the dough less dense – it is leavened. When the dough is baked most of the ethanol evaporates and the carbon dioxide bubbles give the bread a light texture, which makes it more appetizing.

TRADITIONAL FOOD PRESERVATIVES

There are many ways of preserving foods. All of them work by preventing the growth of microbes in the food. Several traditional methods involve adding chemicals to the food.

1. Acids
Vinegar, containing ethanoic acid, can be used to preserve foods, because microbes cannot grow at low pH.
Yoghurt is made when bacteria convert lactose in milk to lactic acid. This reduces the pH of the milk, preventing the growth of most microbes and therefore preserving the milk as yoghurt.

2. Salt
If salt, usually sodium chloride, is added to foods, to create a high salt concentration, microbes cannot grow in the food and it is preserved. Any microbes in the food are killed because water is drawn out of them by osmosis.

3. Sugar
Honey is naturally preserved, because of its high sugar content. If honey is properly ripened, no microbes can grow in it, because the sugar is so concentrated. As with salt, water is drawn, by osmosis, out of any microbes in honey, killing them. If enough sugar is added to a food then it is also preserved. Jam is an example of a food preserved by high sugar concentrations.

PRODUCTION OF SOY SAUCE

The traditional method of producing soy sauce takes about six months and involves a fungus, *Aspergillus*.

1. Soya beans are cooked and are mixed with ground roasted wheat grains.

2. The mixture is inoculated with *Aspergillus*, which grows rapidly over the soya and wheat.

3. After a few days, the mixture is transferred to vats and salt solution is added.

4. The mixture is left in the vats for about six months, during which time the *Aspergillus* ferments the starch and proteins into alcohol, organic acids, sugars and amino acids.

5. Soy sauce is the liquid produced by pressing the mixture extracted from the vats at the end of the fermentation. It is a complex combination of sweet, salty, sour and unami flavours. Its traditional use is in oriental cooking, but it is now used widely throughout the world.

FOOD POISONING

Some microbes that are found in food produce toxins, which are harmful to human health. This is called food poisoning. One of the commonest forms is caused by certain strains of *Staphylococcus aureus*.

Symptoms
If food containing the toxin is eaten, nausea, vomiting and diarrhoea develop with a few hours.

Method of transmission
If food is contaminated with the pathogenic strains of *S. aureus* during handling and the food is stored above 4°C, the bacteria multiply and produce harmful toxins. A wide variety of foods can carry the bacteria and therefore toxins: poultry, meat, eggs, salads, puddings, sauces and bakery products containing cream.

Treatment
The main aim of treatment is to replace substances lost in diarrhoea. Oral rehydration fluids are used, which are dilute solutions of mineral ions, including sodium and chloride, together with a little sugar and some flavouring to make it palatable. Intravenous fluids are only given when vomiting prevents rehydration. Antibiotics are not normally used as the body clears the infection without them.

SOURCES OF ENERGY AND CARBON

Microbes, especially prokaryotes, are much more varied in their metabolism than larger organisms. They can be divided into groups according to their sources of two essential things – energy and carbon. The table (right) summarizes the differences.

Photoautotrophs – organisms that use light energy to generate ATP and to produce organic compounds from inorganic substances.
Example – *Anabaena* (shown in the diagram below)

Photoheterotrophs – organisms that use light energy to generate ATP and that use organic compounds made by other organisms.
Example – *Rhodospirillum* (a purple bacterium)

Chemoautotrophs – organisms that use energy from chemical reactions to generate ATP and that produce organic compounds from inorganic substances.
Example – *Nitrobacter* (a nitrifying bacterium)

Chemoheterotrophs – organisms that use energy from chemical reactions to generate ATP and that use organic compounds made by other organisms.
Example – *Lactobacillus* (a bacterium used to make yogurt)
Plants are almost all photoautotrophs and most animals are chemoheterotrophs.

	SOURCE OF CARBON	
	Inorganic -CO_2	Organic compounds
Light	Photoautotrophs	Photoheterotrophs
Chemical	Chemoautotrophs	Chemoheterotrophs

(SOURCE OF ENERGY)

Structure of a cyanobacterium

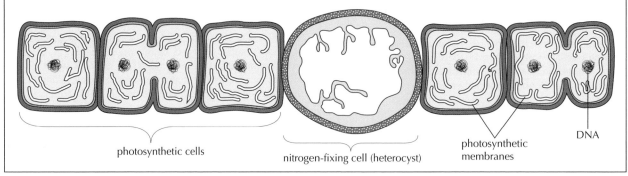

photosynthetic cells   nitrogen-fixing cell (heterocyst)   photosynthetic membranes   DNA

USE OF BACTERIA IN BIOREMEDIATION

Microbes can sometimes be used to clean up pollution of soil or water. This is called bioremediation.
The bacteria may break the pollutant down into harmless substances or may remove it from the environment by concentrating it within the microbial cells.
Usually the bacterium is already present in the environment, but it may need to be stimulated by applications of nutrients, such as nitrate or phosphate fertilizer. Four examples of bioremediation are given here.

1. Oil spills in water

Crude oil contains a wide variety of chemical compounds, many of which have very harmful effects when released into the environment. Many different microbes are able to oxidize hydrocarbons in oil spills, and so help in bioremediation. The numbers of these microbes increase naturally on the surface of the oil, but even so, it can take months for most of a spillage to disappear and some hydrocarbons are very resistant to microbial decomposition. Application of inorganic fertilizer can speed up the process of bioremediation.

2. Selenium pollution

Compounds of the metal selenium sometimes pollute soils or water. Various bacteria absorb selenate ions (SeO_4^{2-} and SeO_3^{2-}) and oxidize them to metallic selenium, which is much less toxic.

3. Pesticide pollution

A wide variety of chemical substances have been developed for use as pesticides. When these substances are used, residues may be left in the soil or elsewhere in the environment. These residues gradually disappear but the time taken for this to happen varies from a few days up to four or more years. This depends on how easy it is for microbes to break down the pesticide.

4. Solvent pollution

Chlorinated solvents, for example chloroform, are often found as pollutants of groundwater – water percolating through soil and rock. There are a few groups of bacteria that dechlorinate these solvents in anaerobic conditions, converting them into much less toxic substances.

Microbes and disease

LIFE CYCLE OF THE INFLUENZA VIRUS

Influenza is caused by an enveloped virus, with single-stranded RNA as its genetic material.

- It binds to glycoproteins on the surface of the cells in the lining of the upper respiratory tract.
- It is then taken into these cells by endocytosis.
- Once inside the host cells, the viral RNA is replicated and capsid proteins are synthesized using the ribosomes of the host cell.
- New influenza viruses are assembled from the RNA and proteins.
- The host cell is burst open. This is called lysis. The influenza viruses are released, enveloped in membrane from the host cell's plasma membrane.
- The viruses that have been released go on to invade other host cells, spreading the infection.

This type of life cycle, where a virus takes over a host cell, uses it to reproduce and then bursts it open and kills it, is called a **lytic life cycle**.

ENDOTOXINS AND EXOTOXINS

Many pathogens harm their hosts by producing toxins.

In some cases the toxin is a protein that is released by the pathogen – this is an **exotoxin**. Many exotoxins are highly toxic or even fatal, for example, the bacterium that causes cholera releases a protein that perforates the membranes of cells in the intestine. This causes loss of fluid from the wall of the intestine and extremely severe diarrhoea. Exotoxins can move through the body of the host and cause damage away from the area of infection.

Another type of toxin is present in the outer membrane of gram-negative bacteria. The toxic part of the membrane is lipopolysaccharide and because it is part of the structure of the bacterium, it is an **endotoxin**. The toxicity of endotoxins is not great, but they cause fever and aches, which exotoxins usually do not.

ACTION OF ANTIBIOTICS

Antibiotics are chemical substances produced by microbes that kill or inhibit the growth of other microbes. Their discovery and use is one of the triumphs of modern medicine, revolutionizing the treatment of bacterial diseases. Antibiotics all interfere with some aspect of microbial metabolism. Most of them act against bacteria, by one of these mechanisms:

1. **Inhibiting of cell wall synthesis** – penicillin and some other antibiotics inhibit enzymes that are involved in the synthesis of the bacterial cell wall

2. **Inhibiting protein synthesis** – erythromycin, streptomycin and some other antibiotics block one of the stages in bacterial protein synthesis

3. **Inhibiting nucleic acid synthesis** – rifampin and some other antibiotics block the synthesis of RNA by RNA polymerase in bacteria

Antibiotics can safely be ingested because these processes are sufficiently different in human cells for them not to be blocked.

TYPES OF BACTERIAL INFECTION

There are two main types of bacterial infection:

1. **Extracellular bacterial infection**
Some pathogenic bacteria invade the body and remain in the intercellular spaces, using the nutrients there.
Example: *Streptococcus*
This group of bacteria most commonly infects the upper respiratory tract. *Streptococcus* cells sometimes form an outer covering, called a capsule, which helps them to resist the antibodies in human tissues.

2. **Intracellular bacterial infection**
Some pathogenic bacteria invade the body of the host and enter its cells, relying on the metabolism of the host cells for some processes.
Example: *Chlamydia*
Small dense *Chlamydia* cells are able to survive outside host cells, but not grow or divide. When they make contact with a host cell, they are taken in by endocytosis. Once inside they change into larger active cells, which use ATP and other substances produced by the host, for growth and reproduction. Eventually these active cells become smaller and denser and are released. They may then be dispersed and enter other host cells.

CONTROLLING MICROBIAL GROWTH

1. **Irradiation** – ionizing radiation, for example gamma radiation, can be used to kill microbes in food, including pathogenic bacteria and those that cause food spoilage. Free radicals formed by the irradiation may alter flavour, but not as much as with heating. Some bacteria can survive irradiation e.g. *Clostridium botulinum*. Irradiated foods do not become radioactive, but some consumers are still reluctant to buy them.

2. **Pasteurization** – milk can contain pathogens, including the bacteria that cause tuberculosis. Pasteurization kills all pathogens and most bacteria causing decay. A typical method involves heating the milk to at least 72°C for 15 seconds, followed by rapid cooling. Longer periods of heating or higher temperatures sterilize the milk, preventing decay more effectively, but altering the flavour of the milk, so pasteurization is often preferred to sterilization.

3. **Antiseptics** – chemical substances that kill or prevent the growth of bacteria on the skin or in wounds, helping to prevent infection.
Antiseptics can be used on the surface of living tissues because they are not very toxic to tissue and there is little or no absorption. However, they would be harmful if taken internally. They cannot therefore be used in foods, and they would also taste unpleasant.

4. **Disinfectants** – chemical substances that kill or prevent the growth of microbes on non-living surfaces. They can be used to sterilize medical equipment, surfaces used in food preparation and many other places where microbial growth must be prevented.
Effective disinfectants are highly toxic to microbes, but the disadvantage of this is that they are too toxic to be used on or in living tissues, or in foods.

Ⓗ Epidemiology

EPIDEMICS AND PANDEMICS

Epidemiology is the study of the occurrence, distribution and control of diseases.

There is an **epidemic** when the number of cases of a disease in a region is unusually high. There is a **pandemic** when an epidemic has spread very widely, to affect a large geographic area, such as a continent.

Example: the Asian flu pandemic of 1957

Occurrence: pandemics of influenza occur irregularly, but usually at intervals of several decades. An epidemic began in February 1957, spread into a pandemic and reached its peak in October 1957, with 22 million new cases in two weeks, declining from then onwards.

Distribution: the new strain of influenza virus that caused the pandemic first appeared in Mainland China, spreading to Hong Kong and then throughout the world by air and sea routes.

Control: clearly, there was no effective control of the pandemic in 1957. When new strains of influenza virus appear, vaccines are still not immediately available, because development and manufacture takes months. There were no effective antiviral drugs available and there are still no drugs that are as effective as antibiotics for bacterial diseases. If a dangerous new strain of influenza is identified early enough, attempts can be made to prevent its spread by isolating all infected people, but this did not happen in 1957.

MALARIA

300–500 million people per year become ill as a result of malaria, with more than a million deaths, making it one of the world's most devastating diseases.

Cause

A protozoan parasite called *Plasmodium* is the cause. After entering the body, it first invades and reproduces inside liver cells and then changes into a different form, which targets red blood cells. In the most severe form of malaria, the parasite follows a 48-hour cycle of invading red blood cells, growing and reproducing inside them and bursting out into the blood plasma.

Effects

The worst symptoms occur while the parasites are circulating in the blood plasma: fever, shivering, sweating, headache, general body pain and stomach upsets. In severe cases the attacks can become progressively more serious. Capillaries are blocked by parasitized red blood cells and burst causing anemia and widespread damage to organs including the brain. Death may then follow.

Transmission

The parasite cannot by itself get from the body of one human host to another. It uses an insect vector – female *Anopheles* mosquitoes, which feed on human blood. If the mosquitoes ingest blood from a person with malaria, the malarial parasites survive and reproduce inside the stomach and then spread to the salivary glands. When the mosquito next feeds, usually on a different person, it first injects saliva that contains *Plasmodium* into the person, infecting them with malaria if they do not already have it.

TRANSMISSION OF PATHOGENS

One of the main problems in the life of a pathogen is how to reach a new host and gain entry to the body. There are various possible methods.

- **Contact** – contagious diseases are transmitted when an uninfected person touches an infected person as the pathogen can enter the body through the skin.
- **Cuts** – pathogens enter the body when the skin is cut or punctured by any object that is contaminated with pathogens.
- **Droplets** – diseases of the ventilation system can be transmitted when an infected person coughs or sneezes out droplets containing pathogens, which are breathed in by an uninfected person.
- **Food or water** – pathogens in contaminated food or water enter the body through the soft gut wall.
- **Sexual intercourse** – sexually transmitted diseases gain entry through the soft mucous membranes of the penis and vagina during sexual intercourse.
- **Insects** – blood-sucking insects inject their mouthparts though the skin and can transmit pathogens that are sucked out in the blood of an infected person.

SPONGIFORM ENCEPHALOPATHIES

These are serious, incurable diseases of mammals. The best-known examples are scrapie in sheep, BSE in cattle and Creutzfeld-Jacob disease (CJD) in humans. In each case the tissues of the brain are gradually broken down, giving a spongy appearance and causing premature aging, dementia and eventually death.

Spongiform encephalopathies are infectious, but the nature of the infectious agent is puzzling.
- Enzymes that digest DNA and RNA do not affect it.
- It is very heat stable and is not easily damaged by ionizing radiation.
- It cannot therefore be a living organism.
- It is affected by chemical treatments that denature proteins.

Research has led to a protein of 254 amino acids, now called **prion protein** or PrP. There are two forms of PrP, the normal form, PrP^C, which is found on the surface of neurons and PrP^{SC}, found in diseased brain tissue.
According to the prion hypothesis, PrP^C is converted into PrP^{SC} by a conformational change and PrP^{SC} causes this change. So, if any PrP^{SC} is present in the brain, it will cause more and more to be produced by a sort of positive feedback. Brain cells attempt to digest it using protease, but part of the PrP^{SC} molecule resists digestion and the resulting protein fibrils accumulate in brain cells, presumably causing symptoms of the disease.

Experiments have shown that when experimental animals are inoculated with PrP^{SC} spongiform encephalopathies develop. However, not all observations can be accounted for by the prion hypothesis. It does not explain how rapidly the different forms of the disease progress, including sporadic CJD and variant CJD in humans. No other hypothesis seems plausible though, so research is focusing on modifications to the prion hypothesis.

F1 The graph below shows some of the effects of discharge of raw sewage into a river.

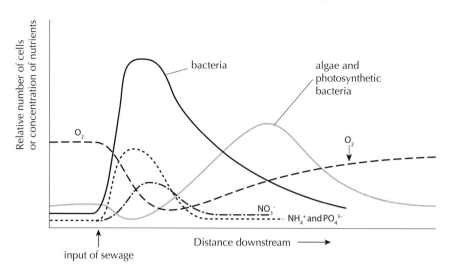

a) Explain the decrease in oxygen concentration downstream of the input of raw sewage.   [2]

b) Explain the increase in ammonia and phosphate concentration downstream of the input of raw sewage.   [2]

c) Explain why the maximum concentration of nitrate is downstream of the maximum concentration of ammonia.   [2]

d) Explain the increase in the numbers of algae and photosynthetic bacteria shown in the graph.   [2]

e) Suggest reasons for the decrease in numbers of bacteria, downstream of the part of the river where there
    is the maximum number of bacteria.   [2]

F2 Reverse transcriptase is an enzyme found only in cells infected by certain viruses.

a) Outline the process catalysed by this enzyme.   [2]

b) (i) State the name of the group of viruses that contain the gene for this enzyme.   [1]

    (ii) State one example of a virus from this group.   [1]

c) Explain briefly why the enzyme is a useful tool for molecular biologists.   [3]

F3 The electron micrograph (right) shows adenoviruses,
    at a magnification of x120 000. Adenoviruses cause the
    common cold in humans.

a) (i) State the name of the outer layer of these viruses,
        visible in the electron micrograph.   [1]

    (ii) State what material is contained inside this outer
         layer of the viruses.   [1]

b) Outline how adenoviruses could be used in
    gene therapy.   [2]

c) Explain whether adenoviruses are intracellular or
    extracellular in their mode of infection.   [2]

Distribution of plants and animals

DISTRIBUTION OF PLANT SPECIES

The distribution of a species is the range of places that it inhabits. The distribution of plants is closely linked to the levels of abiotic factors in the environment. The main abiotic factors are **temperature**, **water**, **light**, **soil pH**, **salinity** and **mineral nutrients**. *Avicennia germinans,* for example, is a tree found in mangrove swamps on the coast of Mexico. It grows where the climate is hot and the soils are waterlogged and anaerobic, with high levels of salinity, a pH close to neutral and high levels of mineral nutrients. Few plants can grow in these conditions, but *Avicennia germinans* thrives.

Sometimes the distribution of a plant species shows what conditions a plant prefers. The figure shows the distribution of *Asperula cynanchica* in Britain and Ireland. It is found in areas with alkaline soils formed from chalk or limestone rock. It is absent from colder northern areas even where the soils are alkaline.

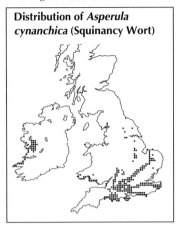

Distribution of *Asperula cynanchica* (Squinancy Wort)

DISTRIBUTION OF ANIMAL SPECIES

The distribution of animal species is affected by both abiotic and biotic factors.

- **Temperature** – external temperatures affect all animals, especially those that do not maintain constant internal body temperatures. Extremes of temperature require special adaptations, so only some species can survive them.
- **Water** – animals vary in the amount of water that they require. Some animals are aquatic and must have water to live in and at the other extreme some animals including desert rats are adapted to survive in arid areas where they are unlikely ever to drink water.
- **Breeding sites** – all species of animals breed at some stage in their life cycle. Many species need a special type of site and can only live in areas where these sites are available. For example, mosquitoes need stagnant water for egg laying.
- **Food supply** – many animal species are adapted to feed on specific foods and can only live in areas where these foods are obtainable. For example, blue whales feed mainly on krill and so congregate in areas of the ocean where krill is abundant.
- **Territory** – some species of animal establish and defend territories, either for feeding or breeding. This tends to give the species an even rather than a clumped distribution. Pairs of tawny owls defend a single territory throughout their adult lives.

RANDOM SAMPLING USING QUADRATS

A sample is a part of a population, part of an area or some other whole thing, chosen to illustrate what the whole population, area or other thing is like. For example, a sample of a population is some individuals in the population but not all of them.

In a random sample, every individual in a population has an equal chance of being selected.

Random sampling of plant populations involves counting numbers in small, randomly located parts of the total area. The sample areas are usually square and are marked out using frames called **quadrats**.

A method for random sampling, using quadrats, is shown in the figure (right).

TRANSECTS AND DISTRIBUTIONS

An alternative to random sampling is to investigate plant or animal distributions along a line marked out across a site. The line is called a **transect**.

Transects are particularly useful when there is a gradient in an abiotic variable. For example, if the soil in a valley is much wetter in the bottom of the valley than up the sides, a transect across the valley can be used to investigate this and the distributions of plant and animal species that are correlated with the variation in soil moisture content.

Transects can be used to investigate plant and animal distributions on seashores. The transect should be laid out at right angles to the high tide and low tide lines, so that it follows the gradient in time of inundation by sea water and time of exposure to air.

Random sampling using quadrats

1. Mark out gridlines along two edges of the area.

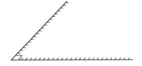

2. Use a calculator or tables to generate two random numbers, to use as co-ordinates and place a quadrat on the ground with its corner at these co-ordinates.

e.g. 14 and 7

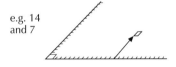

3. Count how many individuals there are inside the quadrat of the plant population being studied. Repeat stages 2 and 3 as many times as possible.

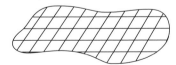

= 5 individuals

4. Measure the total size of the area occupied by the population, in square metres.

5. Calculate the mean number of plants per quadrat. Then calculate the estimated population size using this equation:

$$\text{population size} = \frac{\text{mean number per quadrat} \times \text{total area}}{\text{area of each quadrat}}$$

Niches and interactions

THE NICHE CONCEPT

Studies of the distributions of organisms and of interactions between organisms show that there are many different ways of existing in an ecosystem. The mode of existence of a species in an ecosystem is its ecological niche. The niche includes:

- **Habitat** – where the species lives in the ecosystem.
- **Nutrition** – how the species obtains its food.
- **Relationships** – the interactions with other species in the ecosystem.

If two species have a similar niche, they will compete in the overlapping parts of the niche, for example for breeding sites or for food. Because they do not compete in other ways, they will usually be able to coexist. However, if two species in an ecosystem have exactly the same niche they will compete in all aspects of their life and one of the two species will inevitably prove to be the superior competitor. This species will cause the disappearance of the other species from the ecosystem.

The principle that only one species can occupy a niche in an ecosystem is called the **competitive exclusion principle**.

FUNDAMENTAL AND REALIZED NICHES

The niche that a species could occupy is often smaller than the niche that the species actually occupies. These potential and actual niches are called the **fundamental niche** and the **realized niche** of the species. Differences between the fundamental and realized niches are due to competition.

Other species prevent a species from occupying part of its fundamental niche by out-competing or by excluding it in some other way.

The fundamental niche of a species is its potential mode of existence, given the adaptations of the species.

The realized niche of a species is its actual mode of existence, which results from its adaptations and competition from other species.

Competitive exclusion happens when a species is unable to occupy any part of its fundamental niche in an area, so it has no realized niche in that area.

INTERACTIONS BETWEEN SPECIES

All living organisms are affected by the activities of other living organisms. A situation in which two species affect each other is called an interaction. The table below shows a classification of interactions.

Interaction	Terrestrial example	Marine example
Herbivory – a primary consumer feeding on a plant or other producer. The producer's growth affects food availability for the herbivore.	The beetle *Epitrix atropae* feeds only on leaves of *Atropa belladonna,* often causing severe damage to them. To most other organisms the leaves are highly toxic.	Algae growing on rocks in shallow seas are often heavily grazed. For example, a snail *Lacuna pallida* feeds on the brown seaweed *Fucus serratus* on rocky shores in Europe.
Predation – a consumer feeding on another consumer. The numbers and behaviour of the prey affect the predator.	The Canada lynx is a predator of the Arctic hare. Changes in the numbers of hares (up or down) are followed by similar changes in lynx numbers.	Bonitos feed on anchovetas in the Pacific Ocean west of Peru. When the anchoveta population crashed in the 1970s starving bonitos were found, with completely empty stomachs.
Parasitism – a parasite is an organism that lives on or in a host and obtains food from it. The host is always harmed by the parasite.	The tick *Ixodes scapularis* is a parasite of deer and of white-footed mice in northeast USA. The tick feeds by sucking blood from its hosts and therefore weakens them.	Organisms that cause infectious diseases are all parasites. For example, *Sphingomonas* bacteria cause a disease in elliptical star corals on the Florida reef.
Competition – two species using the same resource compete if the amount of the resource used by each species reduces the amount available to the other species.	Douglas Fir and Western Hemlock grow together in mixed forests in Oregon and other states in northwest USA, competing with each other for light, water and minerals.	Species of coral compete with each other on coral reefs. *Pocillopora damicornis* competes with many other corals, including *Pavona varians*, which benefit when predators feed on *Pocillopora damicornis*.
Mutualism – mutualists are members of different species that live together in a close relationship, from which both benefit.	*Usnea subfloridana* and other lichens consist of a fungus and an alga growing mutualistically. The alga supplies foods made by photosynthesis and the fungus absorbs mineral ions.	The cleaner wrasse is a small fish of warm tropical seas that cleans parasites from the gills and body of larger fish such as reticulate damsel fishes. The cleaner benefits because the parasites that it removes are its food.

Biomass and trophic levels

MEASURING BIOMASS

Ecologists often use a measure called biomass.

Biomass is the total dry mass of organic matter in organisms or ecosystems.

For example, if an ecologist wanted to compare the amounts of organisms in each trophic level in an ecosystem, biomass might be used. Measuring biomass is a destructive technique, so the samples used are as small as possible.

Method

1. Representative samples of all living organisms in the ecosystem are collected, for example from randomly positioned quadrats.
2. The organisms are sorted into trophic levels.
3. The organisms are dried, by being placed in an oven at 60–80°C.
4. The mass of organisms in each trophic level is measured using an electronic balance.
5. Drying and measuring the mass may be repeated to check that samples were completely dry.

DIFFICULTIES WITH TROPHIC LEVELS

Sorting organisms into trophic levels can cause considerable difficulties. This is because many species exist partly in one trophic level and partly in another. The following examples illustrate this.

- Euglena, a unicellular organism found in ponds, has chloroplasts and photosynthesizes, but it also feeds heterorophically by endocytosis.
- Chimpanzees mainly feed on fruit and other plant matter, but they also sometimes eat termites and even larger animals such as monkeys, so they are both first and second consumers.
- Herring are second consumers when they feed on *Calanus* (a copepod) and other first consumers but they are third consumers when they feed on sand eels and other second consumers.
- Oysters (*Ostrea* species) and many other filter feeders consume both ultraplanktonic producers and microplanktonic consumers, so they are first and second consumers. They also consume dead organic matter, so they are also detritivores.

It is difficult to decide into which trophic level these types of organism should be classified. One practical solution is to classify each species according to its main food source.

CONSTRUCTING PYRAMIDS OF ENERGY

Pyramids show the energy flow through each trophic level in an ecosystem. To construct a pyramid of energy, energy flow through each species in the ecosystem must be measured. In each trophic level the energy flow through all species is added up.

The lowest bar of a pyramid of energy is the total amount of energy that flows through the producers in the ecosystem. This is also called gross production.

Gross production is the total amount of organic matter produced by plants in an ecosystem.

Gross production and all the other energy flows in a pyramid are measured in kilojoules of energy per square metre per year ($kJ\ m^{-2}\ year^{-1}$).

Gross production does not have to be measured directly, as it can be calculated from net production and plant respiration.

Net production is the amount of gross production in an ecosystem remaining after subtracting the amount used by plants in respiration.

gross production = plant respiration + net production

Example – an old field community in Michigan, USA.
net production = $20.79 \times 10^3\ kJ\ m^{-2}\ year^{-1}$
plant respiration = $3.68 \times 10^3\ kJ\ m^{-2}\ year^{-1}$
gross production = $(20.79 + 3.68) \times 10^3\ kJ\ m^{-2}\ year^{-1}$
= $24.47 \times 10^3\ kJ\ m^{-2}\ year^{-1}$

The upper bars of a pyramid of energy are the total amounts of energy that flow through the various groups of consumers. This is the amount of energy in the food that the consumers ingest.

The data below was obtained from an Arctic tundra ecosystem on Devon Island in northern Canada.

Trophic level	Energy flow ($kJ\ m^{-2}\ year^{-1}$)
Producers	4925
Primary consumers	24
Secondary consumers	4

This data can be used to construct a pyramid of energy. Each bar of the pyramid should be drawn to the same scale and labelled with the trophic level.

NUMBERS AND BIOMASS OF ORGANISMS IN HIGHER TROPHIC LEVELS

Pyramids of energy show that there are large losses of energy at each trophic level. Reasons for losses of energy are explained on page 41. Losses of energy in ecosystems are accompanied by losses of biomass.

Respiration is an example of a process in which both energy and biomass are lost. When glucose or another respiratory substrate is oxidized in respiration, energy from the glucose is released for use in the cell and is then lost as heat. The mass of the glucose does not disappear – it passes into the carbon dioxide and water that are produced in respiration. When these waste products are excreted, biomass is lost. As a result of respiration and other processes, both energy and biomass are lost at each stage in a food chain.

The energy content per gram of food does not decrease along a food chain. If anything, the food eaten by the higher trophic levels is richer in energy per gram than that eaten by lower trophic levels. However, the total biomass of food available to higher trophic levels is very small. It cannot support large numbers of organisms, especially if these organisms need to be large to overpower their prey. Higher trophic levels therefore usually contain very small numbers of large organisms, with a low total biomass per unit area.

Succession and biomes

ECOLOGICAL SUCCESSION

An ecological succession is a series of changes to an ecosystem, caused by complex interactions between the community of living organisms and the abiotic environment. Two types of succession are recognized:

Primary succession starts in an environment where living organisms have not previously existed, for example a new island, created by volcanic activity.

Secondary succession occurs in areas where an ecosystem is present, but is replaced by other ecosystems, because of a change in conditions. For example, abandoned farmland developing into forest.

During an ecological succession, the community causes the abiotic environment to change. As a result, some species die out and others join the community. Although the community may continue to change in this way for hundreds of years, eventually a stable community develops, called the climax community.

The changes to the abiotic environment during ecological successions vary, but some often occur.

- The amount of organic matter in the soil increases as organic matter released by plants and other organisms accumulates.
- The soil becomes deeper as organic matter helps to bind mineral matter together.
- The soil structure improves as the organic matter content rises, increasing the amount of water that can be retained and the rate at which excess water drains through.
- Soil erosion is reduced by the binding action of the roots of larger plants.
- The amounts of mineral recycling increases, as the soil can hold larger amounts and more minerals are held in the increasing biomass of the community.

BIOMES AND BIOSPHERE

Ecological succession usually stops when a stable ecosystem develops that contains a group of organisms called the climax community. Different types of ecosystem develop in different parts of the world. A type of ecosystem is called a **biome**.

Rainfall and **temperature** are the two main factors that determine what type of ecosystem develops in an area, and therefore what the distribution of biomes around the world is. The climograph below shows the relationship between the levels of these two factors and the types of biome. The characteristics of six major biomes are described below.

The biomes of the world together make up the **biosphere**. It is now well known that the ecosystems and biomes of the world function as one overall ecological system, so the biosphere is the thin layer of interdependent and interrelated ecosystems and biomes that cover the Earth.

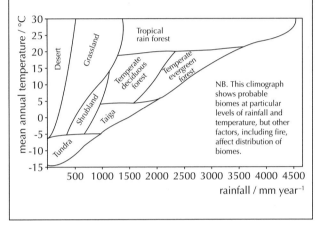

NB. This climograph shows probable biomes at particular levels of rainfall and temperature, but other factors, including fire, affect distribution of biomes.

AN EXAMPLE OF PRIMARY SUCCESSION

On the slopes of Volcan Osorno, in southern Chile, there are large areas of bare volcanic ash, released during recent eruptions of the volcano. Adjacent areas show the stages in an ecological succession.

- Mosses spread over the ash, eventually forming a complete cover.
- Small herbs join the mosses.
- Shrubs, including *Pernettya*, *Eucryphia* and *Embothrium*, enter the community and gradually replace the herbs and mosses.
- Trees, including *Nothofagus*, gradually spread to replace the shrubs with dense forest.

A stage in succession to forest on Volcan Osorno

MAJOR BIOMES OF THE WORLD

Desert	Rainfall very low; warm to very hot days and cold nights.	Very few plants, some storing water and some growing quickly after rain.
Grassland	Rainfall low; warm or hot summers and cold winters.	Dominated by grasses and other herbs that can withstand grazing.
Shrubland	Cool wet winters and hot dry summers, often with fires.	Drought-resistant shrubs dominate, often with evergreen foliage.
Temperate deciduous forest	Moderate rainfall with warm summers and cool winters.	Trees that shed their leaves in the winter dominate with shrubs and herbs beneath.
Tropical rainforest	Rainfall high to very high and hot or very hot in all seasons.	A huge diversity of plants: tall evergreen trees, smaller trees, shrubs and herbs.
Tundra	Very low temperatures; little precipitation mostly as snow.	Very small trees, a few herbs, mosses and lichens are present.

Biodiversity and rainforests

BIODIVERSITY

The word biodiversity was only invented in 1986. It is an abbreviation of 'biological diversity' and encompasses the diversity of ecosystems on Earth, the diversity of species within them, and the genetic diversity of each species. One of the main tasks of ecologists is conservation of the world's biodiversity.

THE SIMPSON DIVERSITY INDEX

It is sometimes useful to have an overall measure of species richness in an ecosystem. The Simpson index is one of the most commonly used.

Method
1. Use a random sampling technique to search for organisms in the ecosystem.
2. Identify each of the organisms found.
3. Count the total number of individuals of each species.
4. Calculate the index (D).

$$D = \frac{N(N-1)}{\Sigma\, n(n-1)}$$

N = total number of organisms
n = number of individuals per species

Example
Organisms were found and identified in the River Enningdalselva in a part of Sweden where some lakes and rivers have been affected by acid rain. Six sites in the river were chosen randomly and kick sampling was used at each site along a 10 m transect. Nets with a 25 cm × 25 cm opening and 0.5 mm mesh were used. The results are shown in the table below.

Group	Species	Name	
Ephemerida	*Dixa* species	Mayfly larva	8
Odonata	*Tipula* species	Dragonfly larva	5
Trichoptera	Species unidentified	Caddisfly larva	4
Plecoptera	*Nemoura variegata*	Stonefly larva	4
Hemiptera	*Gerris* species	Pond skater	3
Isopoda	*Asellus aquaticus*	Water louse	2
Acari	*Arrhenurus* species	Water mite	1
Platyhelminth	*Dendocoelum lact.*	Flatworm	4
Platyhelminth	*Dugesia* species	Flatworm	3
Hirundinea	Species unidentified	Leach	1
Oligochaeta	*Lumbriculides*	Annelid worm	2
Gastropoda	*Lymnaea* species	Snail	4
Bivalvia	*Margaritifer*	Pearl mussel	1

$$D = \frac{42\,(42-1)}{140} = 12.3$$

The high diversity index suggests that the river has not been damaged by acid rain, or any other disturbance. This fits in with observations of a thriving salmon population in the river. If the Simpson diversity index was calculated for another river in the same area, or a river in the same biome elsewhere in the world, the ecological health of these rivers could be compared with River Enningdalselva. This would help to assess whether conservation measures were needed in any of the rivers. It would also allow rivers with high biodiversity to be identified and given appropriate conservation status, for example as nature reserves.

RAINFOREST CONSERVATION

All of the world's biomes must be conserved, but tropical rainforests have been particularly threatened recently and there are many reasons for strenuous efforts to conserve them.

Economic reasons
- New commodities, for example medicines or materials, may be found in rainforest species.
- New crop plants or farm animals could be developed from rainforest species or existing varieties could be improved using their genes.
- Ecotourism could provide considerable income.

Ecological reasons
- Rainforests fix large amounts of carbon dioxide and, without them, the greenhouse effect and global warming would probably be more severe.
- Damage to rainforests can have widespread effects including soil erosion, silting up of rivers, flooding and even changes to weather patterns.

Ethical reasons
- Every species has a right to life, regardless of whether it is useful to humans or not.
- The wildlife of rainforests has cultural importance to the indigenous human populations and it is therefore wrong to destroy it.
- It would be wrong to deprive humans of the future the rich experiences that the Earth's biodiversity provide to us.

Aesthetic reasons
- Rainforests have species in them that are beautiful and give us great enjoyment.
- Painters, writers and composers have been and continue to be inspired by rainforests.

BIOMAGNIFICATION

Some pollutants are absorbed into living organisms and accumulate because they are not efficiently excreted. When a predator consumes prey containing the pollutant and absorbs it, the level in the body of the predator rises and can reach levels much higher than those in the bodies of its prey. This increase is called biomagnification and it can happen at each stage in the food chain.

Biomagnification is the process by which chemical substances become more concentrated at each trophic level.

Polychlorinated biphenyls (PCBs) are chemicals that were used as insulators in electrical devices and as flame-retardants. It was shown as long ago as 1953 that moderate doses killed experimental rats, but manufacture continued until the 1970s.

PCBs have escaped into the environment and are now detectable throughout the world. They are both persistent and highly toxic. Bioaccumulation factors (BAF) for PCBs vary considerably. Examples are given below.

Pathway	BAF
Soil to earthworm	10
Fish to bird or mammal	90
Water to fish	50 000
Water to shellfish	10 000 000

Impacts of humans on ecosystems

IMPACTS OF ALIEN SPECIES

An alien species is a type of organism that humans have introduced to an area where it does not naturally occur. Alien species are sometimes very invasive and cause considerable ecological damage. For example, the floating fern, *Salvinia molesta*, has damaged many lakes in the tropics and sub-tropics. It grows rapidly, doubling the number of leaves in about two weeks, spreading over the water surface and eliminating native plant species by **interspecific competition**. It has been controlled by introducing another alien species – salvinia weevil (*Cyrtobagus salviniae),* which feeds on the floating fern. This is an example of **biological control**.

Salvinia molesta was deliberately transported around the world as an aquarium or pond plant. Alien species have also been introduced accidentally. For example, three species of rat were introduced to the mainland of New Zealand during the 19th century. They caused many species of bird to disappear from the mainland. This is called **species extinction**. Some of these birds were able to survive on islands that remained free of rats. Until the 1950s, Big South Cape Island in the far south of New Zealand remained rat-free and was a haven for many rare birds. Three types were, by then, found nowhere else: South Island saddleback, Stewart Island snipe and Stead's bush wren.

In the mid-1950s black rats (*Rattus rattus*) reached Big South Cape Island. Their numbers rose exponentially and by 1964 there were huge numbers on the island. They attacked eggs, young birds in nests and even adult birds, which were not behaviourally adapted to resist them. This is an example of alien species causing damage by **predation**.

It became obvious that human intervention was needed to save the three rarest species of bird. Ecologists from the New Zealand Wildlife Service trapped as many of the remaining individuals as they could. Only two Stewart Island snipe were trapped and they died soon after, so this species became extinct. Nine Stead's bush wrens were trapped and transferred to another island that was still rat-free. Unfortunately they failed to breed and gradually died out, so this species also became extinct. Forty-one South Island saddlebacks were caught and transferred to two other rat-free islands. They survived and bred and were eventually distributed to other islands. In the 1980s they were re-introduced to Little Barrier Island after another alien species had been eliminated – wild cats.

The South Island saddleback was the first species of bird to be saved from extinction by human intervention. Its future for the moment seems relatively secure.

South Island Saddleback

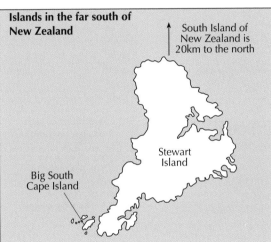

Islands in the far south of New Zealand

South Island of New Zealand is 20km to the north

Stewart Island

Big South Cape Island

Stead's bush wren

OZONE AND ULTRA-VIOLET RADIATION

Ultra-violet radiation has very damaging effects on living organisms and biological productivity.

- It increases mutation rates, by causing damage to DNA.
- It can cause cancers, especially of the skin.
- It causes severe sunburn and cataracts of the eye.
- It reduces photosynthesis rates in plants and algae and so affects food chains.

The amount of damaging ultra-violet radiation reaching the Earth's surface would be much greater without the ozone layer in the atmosphere. Ozone absorbs shortwave radiation, especially ultra-violet. At low altitudes in the atmosphere, the concentration of ozone is usually about 0.01 ppm, but at 20–50km above the Earth's surface, in the stratosphere, ozone is much more concentrated – about 1–10ppm. This is the ozone layer.

Measurements of ozone concentrations in the stratosphere have shown that there has been depletion throughout the world. Since the 1980s an ozone 'hole' has appeared over the Antarctic every year between September and October, which persists for several months.

CFCs are the main cause of ozone depletion. They are chemical compounds manufactured by humans and released into the atmosphere. Ultra-violet light causes CFCs to dissociate and release atoms of chlorine. These chlorine atoms are highly reactive and cause complex reactions in which ozone is converted to oxygen. The reactions form a cycle, with the chlorine atoms being released again, so that they can go on to cause the destruction of more ozone. One chlorine atom can potentially cause the destruction of hundreds of thousands of ozone molecules.

(HL) Conservation

IN-SITU CONSERVATION METHODS

The best place to conserve a species is in its own habitat. This is called *in situ* conservation. Many terrestrial and marine nature reserves have been established for this purpose, but other areas can also be important, including farmland and gardens.

In situ conservation has several advantages.
- Species remain adapted to their habitats.
- Greater genetic diversity can be conserved.
- Animals maintain natural behaviour patterns.
- Species interact with each other, helping to conserve the whole ecosystem.

The size and shape of nature reserves affects their conservation value. The distribution of ecosystems within a nature reserve is also important. These are the **biogeographical features** of a nature reserve. Large nature reserves usually promote conservation of biodiversity more effectively than small ones. The ecology of the edges of ecosystems is different from the central areas, due to edge effects. An example of an edge effect is the egg-laying habits of the cowbird of the western United States. It feeds in open areas on insects disturbed by large grazing mammals, but it lays its eggs in the nests of songbirds, near the edges of forests. Fragmentation of forests has led to a considerable increase in cowbird populations and the nest parasitism due to them, because of the increase in forest edge.

Where a habitat is fragmented, wildlife corridors can be very valuable in allowing organisms to move between different areas, for example tunnels under busy roads.

EX-SITU CONSERVATION METHODS

Despite the advantages of *in situ* conservation, it is not always enough to ensure the survival of a species.

- Some species become so rare that it is not safe to leave them unprotected in the wild.
- Sometimes destruction of a natural habitat makes it essential to remove threatened species from it.

In these situations *ex situ* measures are needed.

1. **Captive breeding** – some or all members of a species are caught and moved to a zoo, where they are encouraged to breed. When numbers are high enough, some are returned to the wild to re-establish a natural population. An example of a species helped by captive breeding is the Hawaiian kestrel.
2. **Botanic gardens** – sites where many different species of plants are cultivated, either in greenhouses or in the open. One of the largest, the Royal Botanic Gardens of Kew, has more than 50000 of the world's 250000 known species in its collection.
3. **Seed banks** – seeds are kept in cold storage at –10ºC to –20ºC. Seeds of most species remain viable for more than a hundred years in these conditions. Other species that are not as long lasting can be germinated and grown to produce replacement seed before viability is lost. The Kew Millennium Seed Bank will eventually hold seed of 25000 endangered species.

ACTIVE MANAGEMENT TECHNIQUES

Some pristine nature reserves can be left in their natural state, but often humans have caused changes and active management is therefore needed to ensure the survival of rare or endangered species. The Hinewai Reserve in the South Island of New Zealand is a good example of limited, but effective, management. Valleys that had been cleared of native forest to become farmland have been allowed to revert to native forest, by secondary succession. Active conservation measures have included the culling of goats. They are an alien species and damage native plants by grazing. Native plants are now re-establishing at Hinewai at an amazing rate.

MONITORING ENVIRONMENTAL CHANGE

Problems in natural ecosystems are detected quickly if there is frequent environmental monitoring. Abiotic factors can be measured directly, but another useful technique is the use of living organisms to detect changes. Indicator species are very useful, as they need particular environmental conditions and therefore show what the conditions in an ecosystem are. Lichens are valuable indicator species because their tolerance of sulphur dioxide varies considerably from the most tolerant to the least tolerant species. Indicator species are also often used to assess pollution levels in aquatic ecosystems. Stonefly, mayfly and caddisfly larvae (below) require unpolluted, well-oxygenated water. Other aquatic species, including chironomid midge larvae, rat-tailed maggot larvae and tubifex worms, indicate low oxygen levels and excessive levels of suspended organic matter, from untreated sewage for example.

Indicator species in aquatic ecosystems

Indicators of high oxygen concentrations	Indicators of low oxygen concentrations
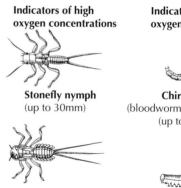	
Stonefly nymph (up to 30mm)	**Chironomid** (bloodworm: a midge larva) (up to 20mm)
Mayfly larva (up to 15mm)	**Rat-tailed maggot larva** (up to 55mm including tube)
Caddisfly larva (up to 30mm)	**Tubifex** (sludge worm) (up to 40mm)

To obtain an overall environmental assessment of a river or other ecosystems, a biotic index can be calculated. There are various methods, which usually involve multiplying the number of individuals of each indicator species by its pollution tolerance rating. An abundance of tolerant species gives a low overall score and an abundance of intolerant species gives a high score.

ESTIMATING ANIMAL POPULATION SIZES

It is usually impossible to count every individual in a population. Instead an accurate estimate is made. Ecologists often need to measure the size of a population. There are many methods for making estimates of population size. The **capture–mark–release–recapture method** is suitable for animals that move around and are difficult to find.

CAPTURE–MARK–RELEASE–RECAPTURE METHOD

1. Capture as many individuals as possible in the area occupied by the animal population, using netting, trapping or careful searching

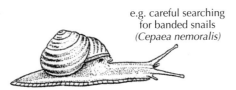

e.g. careful searching for banded snails (*Cepaea nemoralis*)

2. Mark each individual, without making them more visible to predators.

e.g. marking the inside of the snail shell with a dot of non-toxic paint.

3. Release all the marked individuals and allow them to settle back into their habitat.

4. Recapture as many individuals as possible and count how many are marked and how many unmarked.

24 marked

16 unmarked

5. Calculate the estimated population size by using the Lincoln index:

$$\text{population size} = \frac{n_1 \times n_2}{n_3}$$

n_1 = number caught and marked initially
n_2 = total number caught on the second occasion
n_3 = number of marked individuals recaptured

r-STRATEGIES AND *K*-STRATEGIES

Living organisms differ greatly in their life cycles and their patterns of reproduction. As a result, there are different patterns of population growth. Natural selection can cause these characteristics to change, so that they are adapted to the niche of a species. Two extreme patterns of population growth have been defined.

Strategies for an unstable environment

In an unstable environment, life expectancy is very short and few individuals survive long enough to reproduce even once. The population of a species in these environmental conditions is unlikely ever to become large enough for density-dependent factors such as competition to become important. The most successful species use ***r*-strategies**:

- only growing to a small body size, which can be quickly reached
- maturing early, so reproduction happens while still young
- reproducing once only, with all available energy and resources devoted to it
- producing many offspring, with a relatively small body size
- giving offspring little or no parental care.

If all or most offspring survived and reached reproductive maturity, there would be exponential population growth and probably over-population. With *r*-strategists this is very unlikely because the chance of survival of their offspring is so small.

Examples of *r*-strategists

Eschscholtzia californica (Californian poppy)
Lemmus lemmus (lemming)
Clupea harengus (herring)

Strategies for a stable environment

In a stable environment, life expectancy is much longer and many individuals will survive long enough to reproduce repeatedly. The population of a species in these environmental conditions is likely to become large enough for density-dependent factors such as competition to become important. The most successful species use ***K*-strategies**:

- growing to a large body size, which is an advantage in intra-specific competition
- maturing late, with reproduction not beginning until an individual is relatively old
- reproducing more than once and sometimes many times during the extended life-span
- producing few offspring, with a relatively large body size
- giving much parental care to offspring.

Larger numbers of small offspring would make a higher rate of population growth possible. With *K*-strategists, rapid population growth is unlikely ever to continue for long, because it will lead to intense competition. Small offspring, or offspring that are not nurtured by their parents, are unlikely to compete effectively enough to reach adulthood.

Examples of *K*-strategists

Quercus petraea (sessile oak tree)
Loxodonta africana (African elephant)
Dermochelys coriacea (leatherback turtle)

Fish conservation and species extinctions

SUSTAINABLE YIELDS OF FISH

Wild populations of fish are an important food source for many human populations. They are a renewable resource – a resource that need never run out, if it is used in a sustainable way. A renewable resource is constantly replaced or replenished, in the case of fish by them reproducing and growing.

Sustainable use of renewable resource means harvesting at a rate that avoids a decline in the resource. This is particularly important with fish populations. If they are over-exploited and the numbers of adult fish fall below a critical level, spawning fails. The disastrous collapse in the Peruvian anchoveta fishery is an example of this. Industrial scale exploitation of the anchoveta began in 1940 and grew at a rapid rate until 1973, when the annual catch dropped from 12 million tonnes to zero. The fall in anchoveta egg production in the years preceding the population crash is shown in the graph below. An El Nino event was partly responsible, but over-fishing was also a major factor.

Graph showing a collapse in anchoveta egg production

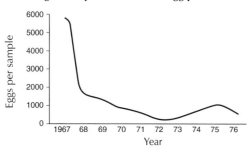

With fisheries, sustainable use means not catching fish faster than the stocks can replenish themselves. The **maximum sustainable yield** is the largest amount that can be harvested without a decline in stocks. One of the aims of research into fisheries is to determine what the maximum sustainable yield of particular fisheries is. International co-operation is then usually needed to ensure that this yield is not exceeded.

INTERNATIONAL CONSERVATION OF FISH

International measures are needed to promote fish conservation because most fish live in international waters, where ships from any country can catch fish. Various measures would help.

- Monitoring of stocks and of reproduction rates.
- Quotas for catches of species with low stocks.
- Closed seasons in which fishing is not allowed, especially during the breeding season.
- Exclusion zones in which fishing is banned.
- Moratoria on catching endangered species.
- Minimum net sizes, so that immature fish are not caught.
- Banning of drift nets, which catch many different species of fish indiscriminately.

Some of these measures have been used already in parts of the world, with limited success. Enforcement is very difficult and relies on a level of international trust and co-operation that is not always seen.

ESTIMATING SIZES OF FISH STOCKS

It is very difficult to estimate the size of commercial fish stocks accurately. This is because fish cannot be seen from above the water surface and many species move around rapidly or are not distributed evenly, so random sampling methods are ineffective.

The usual method of estimating stocks involves collecting data on fish catches. The numbers of fish of each age are counted and an age distribution for the population is obtained. Survivorship curves and spawning rates can then be deduced, from which estimates of the total stock can be made. However there are great uncertainties, for example what proportion of the total population has been caught. Capture–mark–release–recapture methods have been used, with fish marked using internal or external tagging. This method can work well in lakes, but it is less successful in the open sea. By the time the marked fish have mixed back into the overall population by migration, the proportion of marked fish that can ever be recaptured is too small for accurate estimates of the size of the stock.

Other methods of estimation of fish stocks have been used in specific situations. Echo sounders can be used to measure the size of shoals of fish, or even single fish in some cases. The fish must not be swimming too deeply and trawls must also be used for calibration and to check which species of fish has been detected by the echolocation.

None of these methods can estimate stocks with anything approaching certainty and, as a result, disputes between the fishing industry and conservation agencies about stocks are very common.

EXTINCTION OF SPECIES

When the last members of a species die, the species becomes extinct. The rate of species extinctions is probably at an all-time high at the moment, as a result of human activities. There are unfortunately many extinct species from which to select examples for study, including the passenger pigeon and the dodo. The example described here is the Carolina parakeet, *Conuropsis carolinensis*.

These brightly coloured parrots (right) were once common in forests to the east of the Mississippi, from New York to Florida, feeding on seeds of trees and herbs. Clearance of forests reduced their habitat and they started to feed on crops. Farmers killed many of them. Others were caught to obtain feathers, which were used to make fashionable women's clothing. They were also trapped and kept as pets. By 1900 there were no Carolina parakeets in the wild and the last specimen died in Cincinnati Zoo in 1918.

EXAM QUESTIONS ON OPTION G – ECOLOGY AND CONSERVATION

G1 Food chains are difficult to study in natural ecosystems, so a group of ecologists set up communities in culture vessels. They used them to investigate the effects of varying nutrient concentrations. In all of the vessels an aquatic bacterium, *Serratia marcescens,* was present. Three concentrations of the nutrients on which *S. marcescens* feeds were used. In some of the cultures *Colpidium striatum,* a predator of *S. marcescens,* was added. In some of these cultures *Didinium nasutum,* a predator of *C. striatum,* was added. The cultures therefore each had one, two or three trophic levels. The population density of *S. marcescens* at the end of the experiment is shown in the bar chart below.

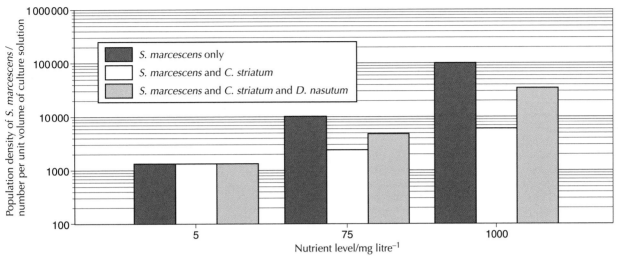

[Source: Kaunzinger, *Nature* (1998), 395, pages 495–496]

a) (i)   Explain the effect of the nutrient concentration on the population density of *S. marcescens.*   [1]

   (ii) Explain the effect of the presence of *C. striatum* on the population density of *S. marcescens.*   [1]

   (iii) Explain the effect of the presence of *D. nasutum* on the population density of *S. marcescens.*   [2]

b) In the culture with the lowest nutrient level *D. nasutum* eventually died out but *C. striatum* survived. Explain the reasons for *D. nasutum* dying out.   [2]

c) Using the results of this investigation, predict a relationship between nutrient levels and length of food chain in natural ecosystems.   [1]

G2 a) Explain how indicator species may be used.   [2]

b) Outline two *ex situ* methods of conservation of endangered species.   [2]

G3 The graph below shows inputs of mercury from the UK to marine waters and flow rates of rivers, between 1990 and 2004, as a percentage of levels in 1990.

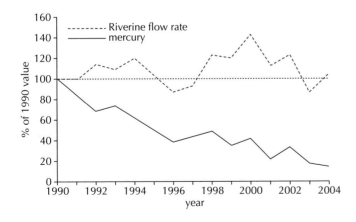

a) State the trend in mercury inputs from the UK to marine waters.   [1]

b) Using the data in the graph, deduce the reasons for fluctuations, from year to year, in mercury inputs.   [2]

c) Biomagnification of mercury can occur in marine ecosystems. Suggest two consequences of biomagnification of mercury in ecosystems.   [2]

Hormonal control

HORMONES

Hormones are chemical messengers, secreted by endocrine glands directly into the blood. The blood carries them to target cells. where they elicit a response. A wide range of chemical substances is used as hormones in humans:

Steroids
e.g. estrogen, progesterone, testosterone

Proteins (peptides)
e.g. insulin, ADH, FSH, LH

Tyrosine derivatives
e.g. thyroxin

MODE OF ACTION OF HORMONES

Hormones do not all work in the same way. There are two main types of mechanism.

1. Steroid hormones enter target cells by passing through the plasma membrane. They bind to receptor proteins in the cytoplasm of target cells, to form a hormone–receptor complex. This complex acts as a regulator of gene transcription, by binding to specific genes. Transcription of some genes is promoted; other genes are inhibited. In this way steroid hormones control whether or not particular enzymes or other proteins are synthesized. They therefore can help to control the activity and development of target cells.

2. Peptide hormones do not enter cells. Instead they bind to receptors in the plasma membrane of target cells. The binding of the hormone causes the release of a secondary messenger inside the cell. The secondary messenger causes a change to the activities of the cell, usually by activating or inhibiting an enzyme.

HYPOTHALAMUS AND PITUITARY GLAND

The hypothalamus is a small part of the brain that links the nervous and endocrine systems. It controls hormone secretion by the pituitary gland located below it (shown in the figure, above right). The anterior and posterior lobes of the pituitary gland are controlled in a different way by the hypothalamus:

Anterior pituitary – neurosecretory cells in the hypothalamus secrete hormones, called releasing hormones, into capillaries in the hypothalamus. These capillaries join to form a blood vessel that lead to the capillaries in the anterior pituitary. This vessel is a portal vein – an unusual type of blood vessel that carries blood directly from one capillary network to another. The releasing hormones stimulate the anterior pituitary to secrete hormones. For example, GnRH stimulates the release of FSH and LH.

Posterior pituitary – neurosecretory cells in the hypothalamus synthesize hormones, pass them via axons to nerve endings in the posterior pituitary and control their secretion. The secretion of ADH is controlled in this way (see right).

Structures of the hypothalamus and pituitary gland

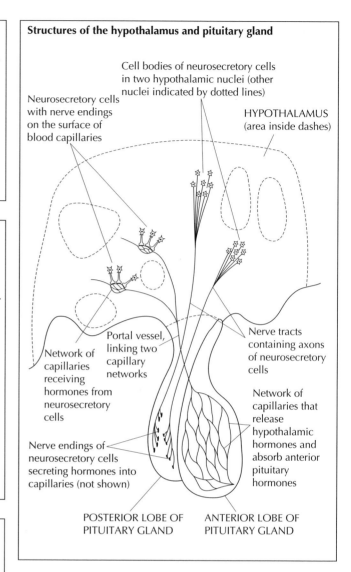

Cell bodies of neurosecretory cells in two hypothalamic nuclei (other nuclei indicated by dotted lines)

Neurosecretory cells with nerve endings on the surface of blood capillaries

HYPOTHALAMUS (area inside dashes)

Portal vessel, linking two capillary networks

Network of capillaries receiving hormones from neurosecretory cells

Nerve tracts containing axons of neurosecretory cells

Network of capillaries that release hypothalamic hormones and absorb anterior pituitary hormones

Nerve endings of neurosecretory cells secreting hormones into capillaries (not shown)

POSTERIOR LOBE OF PITUITARY GLAND

ANTERIOR LOBE OF PITUITARY GLAND

CONTROL OF ADH SECRETION

Neurosecretory cells in the supra-optic nucleus of the hypothalamus synthesize ADH, transport it down their axons and store it in nerve endings in the posterior pituitary gland. Osmoreceptor cells in the hypothalamus monitor the concentration of the blood plasma. If the plasma becomes too concentrated, impulses are passed to the ADH-secreting neurosecretory cells, which convey the impulses to their nerve endings in the posterior pituitary. The impulses stimulate release of ADH into the blood from the stores in the nerve endings. ADH causes a reduction in the concentration of the blood plasma, by stimulating the kidney to produce hypertonic urine (see page 102).

If the osmoreceptor cells detect that the concentration of blood plasma is too low, the neurosecretory cells are not stimulated to release ADH and the blood ADH level rapidly drops, allowing larger volumes of dilute hypotonic urine to be excreted.

Secretion of digestive juices

SUMMARY OF DIGESTION

Food is digested as it passes along the alimentary canal, from the mouth to the anus. Longitudinal and circular muscle fibres in the wall of the alimentary canal contract and relax, squeezing the food and breaking up large solid lumps. Digestive juices, containing enzymes, are mixed with the food. The enzymes digest proteins, nucleic acids, starch and other macromolecules. Digestive juices are secreted by the salivary glands, by glands in the wall of the stomach and by the pancreas. These are all examples of exocrine glands.

Some macromolecules cannot be digested by humans, for example cellulose. The enzyme cellulase digests cellulose, but humans lack the gene that codes for this enzyme, and so cannot make it.

Undigested cellulose is an important part of dietary fibre, which has beneficial effects on the digestive system.

EXOCRINE GLANDS

The secretory cells in an exocrine gland are in a layer that is only one cell thick. The total area of the layer of secretory cells can be very large because of invagination and branching. The digestive juice is released from the cells by exocytosis. It is then discharged from the gland by travelling along ducts. One group of secretory cells, clustered around the end of a duct, is called an **acinus**.
The ducts and acini in part of the pancreas that secretes pancreatic juice are shown below.

Structure of exocrine gland tissue in the pancreas

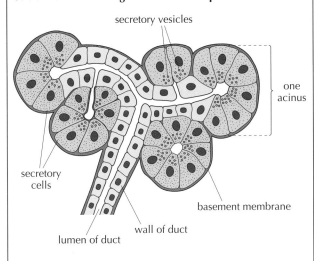

secretory vesicles

one acinus

secretory cells

basement membrane

wall of duct

lumen of duct

EXOCRINE GLAND CELLS

Exocrine gland cells have distinctive features.
- One or two prominent nucleoli inside the nucleus, for production of ribosome subunits.
- An extensive area of rough endoplasmic reticulum, for protein synthesis.
- Golgi apparatuses for processing proteins.
- Many large vesicles, sometimes called secretory granules, for storage of the substances being secreted and transport of them to the plasma membrane. The vesicles are usually densely stained because of the concentration of proteins.
- Mitochondria, to provide ATP for protein synthesis and other cell activities.
The figure (right) is an electron micrograph of a pancreas cell and shows these distinctive features.

CONTROL OF GASTRIC JUICE SECRETION

The control of digestive juice secretion involves both nerves and hormones. The control of gastric juice secretion is described here as an example.
Before food reaches the stomach, gastric juice is already being secreted, as a result of a reflex action. The sight or smell of food stimulates the brain to send nerve impulses to exocrine gland cells in the wall of the stomach. The gland cells start to secrete gastric juice in response.
Much more gastric juice is secreted when food enters the stomach. The food is detected by touch receptors and chemoreceptors in the lining of the stomach and by stretch receptors in the stomach wall. Impulses are sent from these receptors to the brain, which sends more nerve impulses to the exocrine gland cells.
When food is in the stomach, impulses are also sent to endocrine gland cells in the stomach lining that secrete a hormone called gastrin. Gastrin is carried to the exocrine gland cells in the stomach wall, where it stimulates them to increase the secretion of hydrochloric acid. This causes the pH of the food that has entered the stomach to fall to about pH 3.0.

MEMBRANE-BOUND DIGESTIVE ENZYMES

Enzymes secreted by exocrine gland cells become mixed with the food in the alimentary canal and carry out all the initial stages of digestion. However, some of the enzymes that complete the process of digestion work in a different way. They are produced by the wall of the small intestine, but are not secreted. Instead, these enzymes remain in the plasma membranes of cells on the surface of the villi (epithelium cells). The active sites of the enzymes are exposed to the food in the small intestine. They can digest their substrates and the products of digestion can then immediately be absorbed. Epithelium cells tend to be lost from the tips of villi by abrasion, but the membrane-bound enzymes continue to work as they become mixed into the food in the small intestine.

Electron micrograph of an exocrine gland cell in the pancreas (× 6000). The central region of one cell is shown including the nucleus.

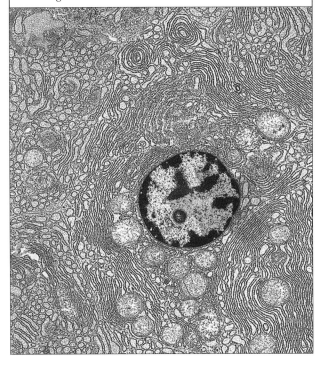

Digestive enzymes

SOURCES OF DIGESTIVE ENZYMES

Food contains many different types of substance that have to be digested before they can be absorbed. Digestion therefore involves many different enzymes, secreted by exocrine glands. The table allows the contents of saliva, gastric juice and pancreatic juice to be compared – there are both similarities and differences.

Digestive juice	Source	Content
saliva	salivary glands	– salivary amylase – mucus
gastric juice	glands in stomach wall	– pepsinogen – hydrochloric acid – mucus
pancreatic juice	pancreas	– pancreatic amylase – pancreatic lipase – phospholipase – trypsinogen – carboxypeptidase – HCO_3^- ions (alkaline)

Pepsin and trypsin are potentially very harmful to the exocrine gland cells that secrete them. They are therefore secreted as inactive precursors, called **pepsinogen** and **trypsinogen**. Pepsinogen is activated by hydrochloric acid, which converts it into pepsin. Different cells in the wall of the stomach secrete pepsinogen and hydrochloric acid (below). Pepsinogen is therefore only activated after it has been secreted. An enzyme, enterokinase, which is secreted by the lining of the small intestine, activates trypsinogen. Activation therefore only happens when trypsinogen enters the small intestine.

Structure of the stomach wall

[Figure: Structure of the stomach wall, with labels – pits, epithelium, interior surface of stomach, cells in neck of gastric gland (secrete mucus), oxyntic cells (secrete hydrochloric acid), peptic cells (secrete pepsinogen), gastric gland]

DIGESTION OF LIPIDS

The digestion of lipids poses special problems, because they are insoluble in water. Foods and the digestive juices added to them are mainly composed of water. In the alimentary canal, lipids in foods melt and form liquid droplets. Because of their insolubility, these droplets tend to coalesce to form larger droplets.

Lipase is water-soluble so it does not enter the lipid droplets, but its active site is hydrophobic (shown on page 68) and hydrolyses lipids on the surface of droplets.

The droplets gradually decrease in size as the lipids on their surface are digested. However, food does not remain in the alimentary canal long enough for large droplets to be digested completely.

Bile helps to overcome this problem. It contains substances called **bile salts**, which are natural detergents. Bile salt molecules have a hydrophobic end and a hydrophilic end. They are therefore attracted to both water and lipids and coat lipid droplets, causing them to break up into smaller droplets. This process is called **emulsification**.

Bile is secreted by the liver and stored in the gall bladder. When it is discharged into the small intestine it emulsifies lipids, which speeds up their digestion, because many small droplets have a larger total surface area, accessible to lipase, than one large droplet of the same volume. With the help of bile, lipids can be completely digested in the small intestine.

THE EFFECTS OF HELICOBACTER PYLORI

Helicobacter pylori is an acid-tolerant bacterium that infects the lining of the stomach. There is evidence that it causes several diseases of the stomach.

1. Stomach ulcers

These are areas of damage to the lining of the stomach. Old medical textbooks state that they are caused by excessive secretion of gastric juice, containing acid. There is now strong evidence that infection of the stomach with *H. pylori* is a more significant factor than gastric acid.

- Antacid treatments may relieve the symptoms of ulcers for a while, but not permanently.
- Antimicrobial treatments that eliminate *H. pylori* infection cure ulcers on a long-term basis.
- *H. pylori* infection is strongly associated with the presence of stomach ulcers.
- Voluntary infection with the bacterium has shown that it can cause gastritis, which often leads to ulceration.
- About half of the *H. pylori* strains isolated from patients with stomach disease produce toxins that cause inflammation – and patients infected with these strains tend to have the most severe ulceration.
- Proteases and other enzymes that are released by *H. pylori* damage the stomach lining.

2. Stomach cancer

Stomach cancer is the growth of tumours in the wall of the stomach. As with stomach ulcers, a far higher percentage of patients with stomach cancer are infected with *H. pylori* than the general population. *H. pylori* infection is associated with reduced vitamin C concentration in gastric juice. This will increase the chance of a tumour forming, but further research is needed to establish a causal link between *H. pylori* infection and stomach cancer.

Absorption of digested foods

STRUCTURE OF THE ILEUM
Digested foods are absorbed in the small intestine, mainly in the latter part, called the **ileum**. The tissue layers of the wall of the ileum are shown in the transverse section below (left). These tissue layers are visible in the light micrograph of the ileum below (right).

Transverse section of ileum

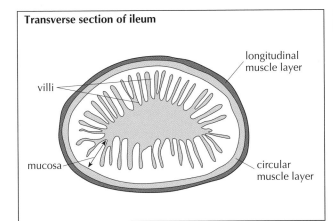

Micrograph of ileum in longitudinal section (× 40)

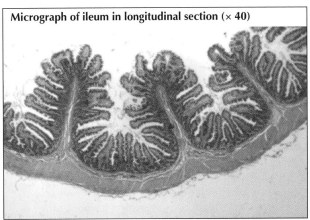

VILLUS EPITHELIUM CELLS
Digested foods are absorbed by villi In the ileum. The structure of a villus is shown on page 47. The outer layer of cells where absorption occurs is the epithelium. The figure (right) is electron micrograph of epithelium cells, showing the structural features that are typical of this cell type. The plasma membranes of adjacent cells are firmly linked together near the free surface by structures called **tight junctions**. These structures prevent molecules from leaking between the epithelium cells. To be absorbed, digested foods have to pass through the plasma membrane of the epithelium cells, and absorption can therefore be carefully controlled. The table below describes the mechanisms used to absorb foods and the structural features used in these mechanisms.

Some materials are not absorbed, including cellulose, lignin, bile pigments, bacteria and abraded intestinal cells. They are therefore egested in the feces.

Micrograph of villus epithelium cells (× 2500)

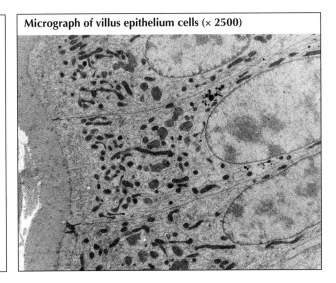

Relationships between structure and function in villus epithelium cells

Structural feature	Function
Microvilli – protrusions of the free surface of the plasma membrane into the lumen of the ileum; about 1 µm long and 0.1 µm wide.	Microvilli greatly increase the surface area of plasma membrane exposed to the digested food in the ileum. This increases the rate of absorption of foods by diffusion. Lipids, and other foods that can pass easily through the hydrophobic centre of the plasma membrane of the epithelium cells, are absorbed by **simple diffusion**. Fructose and some other hydrophilic food substances at a low concentration inside body cells are absorbed by **facilitated diffusion**. There is a steep enough concentration gradient for absorption of these substances by diffusion, but they need assistance to pass through the plasma membrane. Channel proteins help them to cross the hydrophobic centre of the membrane.
Mitochondria – there are many mitochondria scattered through the cytoplasm.	Mitochondria produce the ATP that is needed for absorption of substances by active transport. Pump proteins in the plasma membrane of the microvilli carry out the **active transport**. Glucose, amino acids and mineral ions including sodium, calcium and iron are absorbed in this way.
Pinocytic vesicles – there are many small vesicles, especially near the microvilli.	Pinocytic vesicles are formed by **endocytosis**. Each vesicle contains a small droplet of fluid from the lumen of the ileum. The membranes of these vesicles are formed from the plasma membrane and so contain channels for facilitated diffusion and pumps for active transport. Digested foods can be absorbed from the vesicles into the cytoplasm.

ⓗⓛ Liver

BLOOD FLOW THROUGH THE LIVER

The liver is the largest organ in the human abdomen. It contains huge numbers of cells called hepatocytes, which carry out many vital processes. The liver is supplied with blood by two vessels – the **hepatic portal vein** and the **hepatic artery**. One vessel, the **hepatic vein**, carries blood away. The blood brought by the hepatic portal vein is deoxygenated, because it has already flowed through the wall of the stomach or the intestines. The level of nutrients in this blood varies considerably, depending on the amount of digested food that is being absorbed. One of the main functions of the liver is to regulate levels of nutrients before the blood flows on to the rest of the body. Excessively high levels of glucose and other nutrients would cause damage to the organs of the body, especially the brain.

Inside the liver, the hepatic portal vein divides up into vessels called **sinusoids**. These vessels are wider than normal capillaries, with walls that are more porous. The walls consist of a single layer of very thin cells. There are many pores or gaps between the cells but no basement membrane. Blood flowing along the sinusoids is therefore in close contact with the surrounding hepatocytes.

The hepatic artery supplies the liver with oxygenated blood from the left side of the heart via the aorta. Branches of the hepatic artery join the sinusoids at various points along their length, providing the hepatocytes with the oxygen that they need for aerobic cell respiration.

The sinusoids drain into wider vessels that are branches of the hepatic vein. Blood from the liver is carried by the hepatic vein to the right side of the heart via the inferior vena cava. The figure (below) shows the relationships between blood vessels in liver tissue.

Structure of a sinusoid in the liver

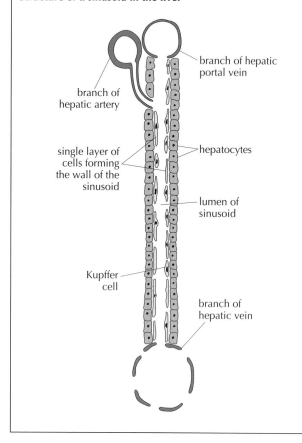

ROLES OF THE LIVER

Nutrient storage and regulation

When certain nutrients are in excess in the blood, hepatocytes absorb and store them, releasing them when they are at too low a level. For example, when the blood glucose level is too high, insulin stimulates hepatocytes to absorb glucose and convert it to glycogen for storage. When the blood glucose level is too low, glucagon stimulates hepatocytes to break down glycogen and release glucose into the blood. Iron, retinol (vitamin A) and calciferol (vitamin D) are also stored in the liver when they are in surplus and released when there is a deficit in the blood.

Breakdown of erythrocytes

Erythrocytes, also called red blood cells, have a fairly short lifespan of about 120 days. The plasma membrane becomes fragile and eventually ruptures, releasing the hemoglobin into the blood plasma. The hemoglobin is absorbed by **phagocytosis**, chiefly in the liver. Some of the cells in the walls of the sinusoids are phagocytic, called **Kupffer cells**. Inside these cells the hemoglobin is split into heme groups and globins. The globins are hydrolysed to amino acids, which are released into the blood. Iron is removed from the heme groups, to leave a yellow coloured substance called bile pigment or bilirubin. The iron and the bile pigment are released into the blood. Much of the iron is carried to bone marrow, to be used in production of hemoglobin for new red blood cells. The bile pigment is used for bile production in the liver.

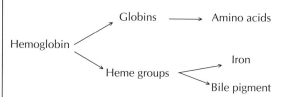

Synthesis of plasma proteins

The rough endoplasmic reticulum of hepatocytes produces 90% of the proteins in blood plasma, including all of the albumin and fibrinogen.

Synthesis of cholesterol

Although some cholesterol is absorbed from food in the intestine, a larger quantity is synthesized each day by hepatocytes.

Detoxification

Hepatocytes absorb toxic substances from blood and convert them by chemical reactions into non-toxic or less toxic substances.

LIVER DAMAGE FROM ALCOHOL ABUSE

Liver cells absorb alcohol and convert it into other substances to detoxify it. Excessive consumption of alcohol therefore damages liver cells more than most other parts of the body. Fatty deposits build up, which can cause hepatitis. Alcoholic hepatitis is inflammation of the liver, often associated with nausea and jaundice. If this is persistent (chronic), for example after ten or more years of heavy drinking, it can cause cirrhosis – normal liver tissue is gradually replaced by scar tissue. Liver cells gradually die and are not replaced, so liver function becomes increasingly poor and eventually death can result from liver failure.

EVENTS OF THE CARDIAC CYCLE

The sequence of actions occurring repeatedly in a beating heart is called the cardiac cycle. The cardiac cycle is described briefly on page 48. The figure below shows the pressure and volume changes in the left atrium, left ventricle and aorta, during two cycles. It also shows electric currents (electrocardiogram) and sounds (phonocardiogram) generated by the beating heart.

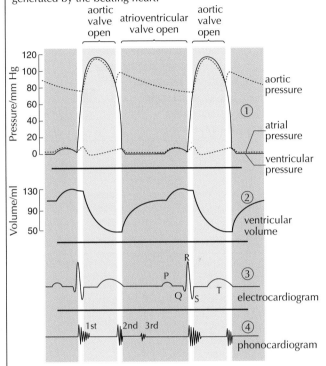

Contraction of the chambers of the heart is called **systole** and relaxation is called **diastole**.

1. Atrial systole

The cardiac cycle begins with the contraction of the wall of the atrium. This happens when the ventricle is already 70% full. The contraction of the atrium pumps more blood into the ventricle, filling it to its maximum capacity before the start of ventricular systole.

2. Ventricular systole

Contraction of the ventricle wall causes a rapid increase in pressure inside the ventricle. This causes the closure of the atrio-ventricular valve, with resulting vibrations in the valve and adjacent walls of the heart. These vibrations are the first heart sound. The pressure in the ventricle rapidly rises above the pressure in the aorta, causing the aortic (semi-lunar) valve to open. Blood can then be pumped from the ventricle into the aorta, raising the aortic blood pressure and decreasing the volume of blood in the ventricle to a minimum. While the ventricle is contracting, the atrium is relaxing and blood enters it from the pulmonary veins.

3. Ventricular diastole

Relaxation of the ventricle wall causes pressure in the ventricle to fall below the pressure in the aorta. The semi-lunar valve therefore closes, with the resulting vibrations that are the cause of the second heart sound. When the pressure in the ventricle falls below the pressure in the atrium, the atrio-ventricular valve opens and blood that has accumulated in the atrium flows into the ventricle causing a rapid rise in ventricular volume. With both the atrium and the ventricle relaxed, blood continues to drain from the pulmonary veins through the atrium into the ventricle until by the end of the cycle it is about 70% full.

CONTROL OF THE HEART BEAT

Heart muscle cells are stimulated to contract by electrical impulses. Interconnections between adjacent cells allow impulses to spread through the wall of the heart, stimulating it to contract. A small region in the wall of the right atrium initiates each impulse (right). This region is called the SA node (sinoatrial node) and acts as the pacemaker of the heart. Impulses initiated by the SA node spread out in all directions through the walls of the atria, but are prevented from spreading directly into the walls of the ventricles by a layer of fibrous tissue. Instead, impulses have to travel to the ventricles via a second node, called the AV node (atrioventricular node). This node is positioned in the wall of the right atrium, close to the junction between the atria and ventricles. Impulses reach the AV node 0.03 seconds after being emitted from the SA node. There is a delay of 0.09 seconds before impulses pass on from the AV node, which gives the atria time to pump blood into the ventricles before the ventricles contract. Impulses are sent from the AV node along two bundles of conducting fibres that pass through the septum between the left and right ventricles, to the base of the heart. Narrower conducting fibres branch out from these bundles and carry impulses to all parts of the walls of the ventricles, causing almost simultaneous contraction throughout the ventricles. The effects of nerves and hormones on the heart beat rate are described on page 48.

Structures involved in the control of the heart beat

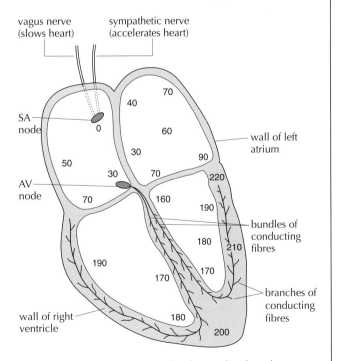

Numbers represent the time taken for impulses from the pacemaker to reach different parts of the heart wall

Coronary heart disease

ATHEROSCLEROSIS

Atherosclerosis is a degenerative disease of large and medium sized arteries. Phagocytes are attracted to sites of damage to the inner lining of the arteries. The phagocytes release growth factors that stimulate the muscle and fibrous tissues in the artery wall to thicken. LDL may penetrate the damaged areas and release cholesterol, which can build up to form large deposits. The growth of wall tissue and accumulation of cholesterol cause the artery wall to bulge inwards, reducing or even preventing the flow of blood. The thickened wall loses its elasticity and calcium salts are sometimes deposited in it, making it hard.

The figure (below) shows a healthy coronary artery and another that shows signs of atherosclerosis.

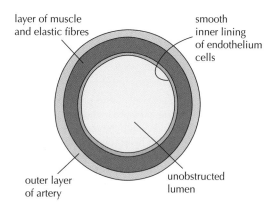

Structure of an artery showing atherosclerosis

CORONARY THROMBOSIS

The rough inner surface of atherosclerotic arteries tends to cause blood clots to form. The formation of clots is called thrombosis.

The wall of the heart is supplied with blood by the coronary arteries. If a blood clot blocks one of these arteries, part of the wall of the heart is deprived of its supply of oxygen. The cells in this part of the wall are unable to respire and so stop contracting. This is either called myocardial infarction or a heart attack. Sometimes small, uncoordinated contractions continue. These are called fibrillations, but they do not pump blood effectively.

FACTORS AFFECTING THE RISK OF CHD

Atherosclerosis and coronary thrombosis are together known as coronary heart disease (CHD). The rates of CHD vary widely between countries.

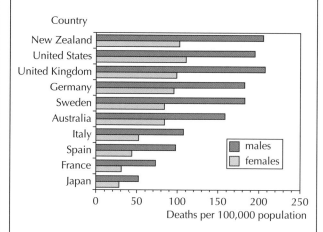

Much research has been done to try to identify factors that increase the risk of CHD. The following factors all increase the statistical risk:
• Increasing age
• Being male rather than female
• Having a family history of CHD

These three factors are not influenced by a person's lifestyle, but some of lifestyle factors that increase the risk are:
• Obesity
• Physical inactivity
• High blood pressure
• Tobacco smoking

The effect of diet is more equivocal. There is some evidence for dietary factors increasing the risk of CHD:
• Trans fat – positively correlated with CHD rates and the data is difficult to explain in any way other than that trans fats cause CHD
• Saturated fat intake – positively correlated with CHD rates in some countries, but evidence of a causal link is lacking.
• Cholesterol intake – reducing dietary cholesterol tends to reduce blood cholesterol levels slightly, and there is a positive correlation between blood cholesterol levels and CHD, but it is a weaker correlation than with saturated fat, and again the causal link is not proven. Cholesterol in blood can be part of both low-density and high-density lipoprotein (LDL and HDL). Whereas high LDL levels are associated with an increased risk of CHD, high HDL levels are associated with a reduced risk. This is because HDL is used to remove cholesterol from tissues.

The levels of LDL, HDL and saturated fats in the blood are not solely due to diet – genetic factors are also important. This may explain why some populations consume large quantities of cholesterol and saturated fats and yet have extremely low CHD rates –the Maasai of Kenya for example.

Finally, there have been claims that some factors reduce the risk of CHD. An example is cis-unsaturated fatty acid intake. These fatty acids are found in olive oil and may explain low CHD rates in Mediterranean countries. However, more evidence is needed before a causal link is established.

Oxygen is transported from the lungs to respiring tissues by hemoglobin in red blood cells. Hemoglobin is a protein that is highly adapted to its function.

OXYGEN DISSOCIATION CURVES

If air with the normal oxygen content is bubbled through a sample of blood, oxygen binds to the hemoglobin until almost all of the hemoglobin molecules have four oxygen molecules bound. The hemoglobin is nearly 100% saturated. If air with a low oxygen content is then bubbled through, some of the oxygen dissociates from the hemoglobin, reducing its percentage saturation. The oxygen content of the air is measured as a **partial pressure**. *Partial pressures are the pressures exerted by each of the gases in a mixture of gases.* The percentage saturation of hemoglobin with oxygen at each partial pressure of oxygen is shown on an oxygen dissociation curve. The figure (below) shows the oxygen dissociation curves of hemoglobin and myoglobin.

Myoglobin is a protein consisting of one globin and one heme group that is used to store oxygen in muscles. The oxygen curve for myoglobin is to the left of the curve for adult hemoglobin because myoglobin has a higher affinity for oxygen. At moderate partial pressures of oxygen, adult hemoglobin releases oxygen and myoglobin binds it. Myoglobin only releases its oxygen when the partial pressure of oxygen in the muscle is very low. The release of oxygen from myoglobin delays the onset of anaerobic respiration in muscles during vigorous exercise.

The dissociation curves for myoglobin and hemoglobin have different shapes. The curve for hemoglobin is S-shaped and that for myoglobin is not. Myoglobin consists of one heme group attached to a globin, whereas hemoglobin has four heme groups, each attached to different globins that interact with each other. As oxygen molecules dissociate from hemoglobin, conformational changes occur, which make it easier for other oxygen molecules to dissociate. Blood containing adult hemoglobin therefore releases large amounts of oxygen over a narrow range of oxygen partial pressures, corresponding to the conditions normally found in respiring tissues.

Oxygen dissociation curves of hemoglobin and myoglobin

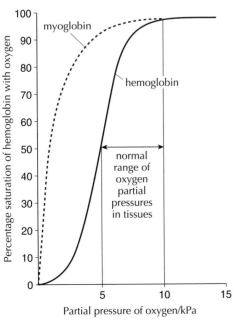

THE BOHR SHIFT

The release of oxygen by hemoglobin in respiring tissues is promoted by an effect called the Bohr shift. Hemoglobin's affinity for oxygen is reduced as the partial pressure of carbon dioxide increases (below). Respiring tissues have high partial pressures of carbon dioxide, so oxygen tends to dissociate. The lungs have lower partial pressures of carbon dioxide, so oxygen tends to bind to hemoglobin.

Effect of CO_2 on the oxygen dissociation curve of hemoglobin

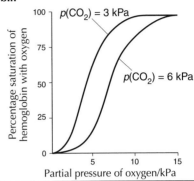

FETAL HEMOGLOBIN

The hemoglobin in the red blood cells of a fetus is slightly different in amino acid sequence from adult hemoglobin. The figure (below) shows that it has greater affinity for oxygen and so, in the placenta, the oxygen that dissociates from adult hemoglobin binds to fetal hemoglobin, which only releases it once it enters the tissues of the fetus.

Oxygen dissociation curves of adult and fetal hemoglobin

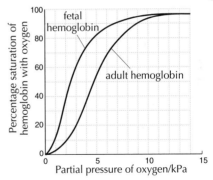

GAS EXCHANGE AT HIGH ALTITUDE

The partial pressure of oxygen at high altitude is lower than at sea level. Hemoglobin may not become fully saturated as it passes through the lungs, so tissues of the body may not be adequately supplied with oxygen. A condition called mountain sickness can develop, with muscular weakness, rapid pulse, nausea and headaches. This can be avoided by ascending gradually to allow the body to acclimatize to high altitude. During acclimatization the ventilation rate increases. Extra red blood cells are produced, increasing the hemoglobin content of the blood. Muscles produce more myoglobin and develop a denser capillary network. These changes help to supply the body with enough oxygen. Some people who are native to high altitude show other adaptations, including a high lung capacity with a large surface area for gas exchange, larger tidal volumes and hemoglobin with an increased affinity for oxygen.

Carbon dioxide transport

Carbon dioxide is produced by aerobic respiration in cells and then either diffuses directly into capillaries or into tissue fluid that is drawn into capillaries. Carbon dioxide is carried by the blood to the lungs in three different ways. A small amount (7%) is carried dissolved in the plasma. The remainder is either converted to hydrogen carbonate ions or binds to hemoglobin.

CONVERSION TO HYDROGEN CARBONATE IONS

Carbon dioxide can be converted into hydrogen carbonate ions within a fraction of a second of entering the blood. About 70% of carbon dioxide is carried in this way. After diffusing into red blood cells, carbon dioxide combines with water to form carbonic acid. This reaction is catalysed by carbonic anhydrase. Carbonic acid rapidly dissociates into hydrogen carbonate and hydrogen ions. The hydrogen carbonate ions move out of the red blood cells by facilitated diffusion. A carrier protein is used that simultaneously moves a chloride ion into the red blood cell. This is called the **chloride shift** and prevents the balance of charges across the membrane from being altered. The figure (below) shows the reactions that produce hydrogen carbonate ions and the chloride shift.
The hydrogen ions that dissociate from carbonic acid bind to hemoglobin in the red blood cells, preventing an excessive change in pH. This is called pH buffering. Plasma proteins also act as pH buffers in blood.

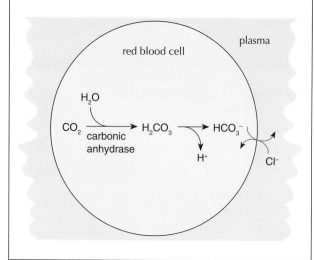

BINDING OF CARBON DIOXIDE TO HEMOGLOBIN

In respiring tissues, carbon dioxide binds reversibly to hemoglobin, to form carbaminohemoglobin.
In the lungs, **carbaminohemoglobin** dissociates and the carbon dioxide is released. Between 15% and 25% of carbon dioxide is carried in this way. The binding of carbon dioxide and hydrogen ions to hemoglobin lowers its affinity for oxygen. This causes the Bohr shift (page 169).

THE EFFECT OF EXERCISE ON VENTILATION

During vigorous exercise, the energy demands of the body can increase by over ten times. The rate of aerobic respiration in muscles rises so there is an increase in the amount of CO_2 entering the blood and the concentration rises. This reduces the pH of the blood and is rapidly detected by cells in the walls of arteries, which monitor blood pH and concentrations of oxygen and carbon dioxide in the blood. These cells are called **chemosensors**. The chemosensors send nerve impulses to the parts of the medulla of the brain that control the ventilation rate, called the **breathing centres**. The breathing centres also monitor blood pH and carbon dioxide concentration. If the concentration of carbon dioxide in the blood rises and the blood pH falls below its normal level of pH 7.4, the breathing centres increase the rate of inspiration and expiration. This is done by sending nerve impulses to the diaphragm and intercostal muscles, causing them to increase the rate at which they contract and relax. The increase in the ventilation rate helps to remove from the body the CO_2 produced in aerobic cell respiration. It also helps to increase the rate of oxygen uptake, which allows aerobic cell respiration to continue in the muscles and it helps to repay the oxygen debt after anaerobic cell respiration.
After exercise, the level of CO_2 in the blood falls, the pH of the blood rises and the breathing centres cause the ventilation rate to decrease.
The figure (below) shows the relationship between blood pH, partial pressure of carbon dioxide in blood and ventilation rate.

Effect of varying blood pH and CO_2 level on the ventilation rate

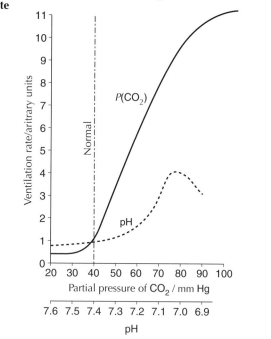

ASTHMA

During asthma attacks the muscles in the wall of the bronchi contract excessively, narrowing the bronchi. Ventilation is a struggle and gas exchange is reduced.
Asthma is an allergic reaction, often to house dust mites, but sometimes also to pollen, pets and some fungi. According to a recent theory, living in very clean homes increases the risk. Without enough pathogens to fight, the immune system starts to react against harmless substances, causing allergies to develop.

EXAM QUESTIONS ON OPTION H – FURTHER HUMAN PHYSIOLOGY

H1 The bacterium *Helicobacter pylori* infects the lining of the stomach. A survey was done using patients who had complained of pain or discomfort in their digestive system. The lining of their oesophagus, stomach and duodenum (upper part of the small intestine) was examined using an endoscope and the patients' blood was tested for the presence of antibodies against *H. pylori*.

The table below show the results of the survey.

Endoscopy finding	Antibodies against *H. pylori* present (number of patients)	Antibodies against *H. pylori* absent (number of patients)
Normal	51	82
Oesophagus inflamed	11	25
Stomach ulcer	15	2
Stomach cancer	5	0
Duodenum inflamed	15	2
Duodenal ulcer	24	1

a) Explain why the researchers tested for antibodies against *H. pylori* in the blood of the patients. [2]

b) Discuss the evidence from the survey results for *H. pylori* as a cause of stomach ulcers and stomach cancer. [3]

c) (i) Compare the results for inflammation of the oesophagus and the duodenum. [2]

   (ii) Suggest a reason for the difference in the results. [2]

H2 a) Outline how the atria of the heart are stimulated to contract. [2]

   b) Explain the origin of the heart sounds. [2]

H3 V_E is the total volume of air expired from the lungs per minute. The graph below shows the relationship between V_E and the carbon dioxide content of the inspired air.

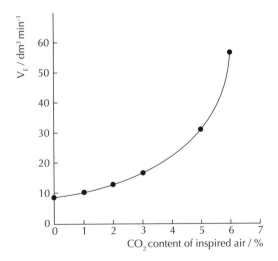

a) Outline the relationship between the carbon dioxide content of inspired air and V_E. [2]

b) Explain how the carbon dioxide content of inspired air can affect V_E. [3]

c) Predict the effect on V_E of increasing the carbon dioxide concentration of inspired air above 7%. [2]

During the IB Biology course, your teacher will help you to improve your skills in planning and performing investigations. When your teacher thinks that you are ready, your skills will be assessed. This will be done during lessons at your school, so it is called internal assessment (IA). Exams that are sent off to an examiner are called external assessment. In IB Biology, 24% of the marks are for IA.

Although your teacher will help you as much as possible, you cannot be given higher marks than you deserve, because samples of work from your school are checked to see whether they have been marked to the right standard. This is called external moderation. You must therefore demonstrate high levels of skill to gain high marks in IA.

Five skills are assessed in IA – these are called the IA criteria. They are Design; Data collection and processing; Conclusion and evaluation; Manipulative skills; Personal skills. Each of these criteria is divided into three aspects. For each aspect, your teacher will decide whether you deserve a mark of 2 (for meeting this aspect of the criterion fully), 1 (for meeting it partially) or 0 (for not meeting it at all). The maximum mark for a criterion is therefore 6. The first three criteria are assessed on at least two occasions and your highest two scores are selected. There will be one overall assessment of your manipulative skills, to reflect the skill level that you have reached during the course. There will be one overall assessment of personal skills. This will be done during the Group 4 project – the co-operative science project that you participate in, at some stage during the course. The maximum possible mark for IA is therefore 48 (2 x 6 for the first three criteria and 1 x 6 for the last two criteria).

The comments below explain how to improve your level of skill in each of the five criteria.

IA criterion	Guidance
Design	1. The first aspect of design is choosing a problem or research question to investigate. You are expected to pick one factor to vary in an experiment. This is the independent variable and it is expected to affect the level of another variable, called the dependent variable. You are also expected to identify other variables that could affect the independent variable and therefore must be kept constant. These are called controlled variables. For example, if you deliberately vary temperature, the activity of an enzyme will vary, but pH, enzyme concentration and substrate concentration must be kept constant.
	2. The second aspect of design is planning how to manipulate the level of the independent variable, and how to keep the levels of all the controlled variables constant.
	3. The third aspect of design is the detailed planning of the method. This includes deciding on the range of the independent variable and also the method for precise and accurate measurement of the dependent variable. You must also decide how many measurements you need to make. One measurement may sometimes be enough, but you can only calculate the standard deviation if you repeat measurements to give a sample of at least five for each level of the independent variable.
Data collection and processing	1. The first aspect is recording results of an experiment – the results are called raw data. Usually the raw data is quantitative – measurements with S.I. units. Record these measurements as accurately as possible using a results table that you have produced yourself, either handwritten, or word-processed. Show every result that you obtained, not just the mean results. All the results should be given to the same number of decimal places. The column headings on results tables should show both the quantity being measured and the S.I. units. When you are being assessed for this aspect, you should show the size of the uncertainty with each result. For example, the time given by a stopwatch might be plus or minus 10 milliseconds, or with a ruler it might only be possible to measure lengths to plus or minus 0.5 mm.
	2. To satisfy the second aspect you must process the raw data in some way. This might involve calculating the mean, or a percentage. Processing of raw data makes it suitable for plotting on a graph. An example of this is calculating the percentage mass change of samples of potato, placed in different salt solutions.
	3. The third aspect of this criterion involves presenting the processed data as a graph or other appropriate chart, chosen by you. Remember these rules for graphs: • put the independent variable on the x-axis and the dependent variable on the y-axis • choose an appropriate scale for the x-axis and the y-axis so that the graph is a suitable size • label both the axes and remember to include units (for example, mass/grams) • plot the data points accurately and join them with a best-fit curve or straight line • if you have calculated mean results, plot these rather than individual results • error bars are could be added to show the range of uncertainty above and below each point – there are several ways of doing this, but the most usual is to show the standard error. (Standard error = standard deviation divided by the square root of the sample size).

Conclusion and evaluation	1. The first aspect of this criterion is drawing valid conclusions from the data that you obtained. These questions can be used as prompts. What trends or patterns are shown by the data? What relationship is there between the independent and dependent variables? Your graph should give you the evidence for the relationship. If you calculated mean results, is there a significant difference between the means? What is the explanation for the observed relationships or differences between means? 2. The next task is commenting on the design of the investigation and the experimental methods used. You should list all of the weaknesses, including measurements not being precise enough, results being inaccurate because of errors, or allocation of time to the various parts of the investigation not being appropriate. For each weakness in the investigation, you must assess how significantly it affected the results. The process of identifying weaknesses and assessing their impact is called evaluation. 3. The third aspect of this criterion is explaining what could be done to improve the investigation, if it was done again. You must be specific – for instance, it is not enough to say that more precise measurements should be made; you must explain how more precise measurements could be made.
Manipulative skills	1. By the time that your teacher assesses your manipulative skills, you must have shown that you can follow instructions safely and accurately, even with complicated laboratory practical work. Be sensible about asking for help from your teacher. Try to work out what to do yourself. But, if you have not been given full enough instructions or are worried about the safety of the procedure, ask for help. 2. During the course, your teacher should give the opportunity to learn how to use a wide range of techniques and equipment. By the time your skills are assessed you should be competent in all these techniques. Work in a careful and systematic way – arrange your apparatus tidily and do not waste time, but work without rushing. 3. You should always know about any potential risks in the procedure that you are following. You should never put yourself or others in the laboratory at risk of accident or injury.
Personal skills	1. You will be assessed on your approach to the co-operative science project (Group 4 project). You are expected to show self-motivation and follow through the project until it is completed. Obviously, you will not satisfy this aspect if your teachers have to encourage you to persevere with the project! 2. Scientists often work in teams, so the ability to co-operate with others is important. During the Group 4 project, you must show that you can collaborate with others, by communicating effectively and working co-operatively. You should ask yourself these questions: Are you only interested in your own views or do you ask for the views of others? To satisfy this aspect you must exchange ideas with others and help to combine them so that the team completes a task more effectively than any one individual could. 3. You may be asked to complete a self-evaluation form, to allow this aspect of personal skill to be assessed. You are expected to show a realistic awareness of your own strengths and weaknesses. You are also expected to explain what you have learned from the Group 4 project.

Many IB students choose a biological research question for their extended essay. There are unlimited opportunities for novel and interesting work because of the diversity of life. Many excellent Biology extended essays are written each year. Every essay is an individual effort and there is no formula for writing the perfect essay. The steps shown below are intended to help you to avoid some common faults, without preventing you from writing the essay that you want to write.

While you are working on the essay, your most important resource will be the teacher who is supervising you. If you need help at any stage, fix a time to talk things over. You should make sure that your supervisor always knows what you are doing – discuss how things are going as frequently as possible. If you don't, you could waste a lot of time on an unproductive approach to the work. Remember two important maxims: 'things take time' and 'if something can go wrong it will'. Assume from the start that you'll have to do a second run of any experiments or observations and allow time for this. Start work as soon as possible and then you will have time to produce the finest essay that you can. You can also earn extra points towards your diploma.

Planning and data collection

Choose a suitable topic	Pick the field of biology that interests you most and gradually narrow down to one small section of it. You must choose a truly biological topic – one that involves living organisms and interactions between them. It must be a topic in which you can have a personal input – this isn't easy with some topics, for example diagnosis and treatment of diseases, so these are best avoided.
Choose an approach	There are two main types of approach. 1. Doing experiments/making careful observations. 2. Searching in books, journals or on the internet for relevant data. Most of the best essays combine both approaches. If you cannot design and do experiments in your chosen topic, reconsider your choice of topic!
Do some preliminary work	Try out some experiments – this should allow you to find out whether your approach is likely to be successful. Avoid experiments that cause unnecessary risks, suffering to animals or environmental damage. Do some background reading and take careful notes of important relevant information. Preliminary work should get you thinking and asking questions about your topic.
Formulate a research question	This should be a question worth asking – not one with an obvious answer. It should be narrow enough to be fully answered in a 4000-word essay, based on 40 hours' work. It is best stated as a question, which can be used to develop a hypothesis – a prediction that you are going to test. It is absolutely vital at this stage to talk to your main helper – your teacher.
Plan your methods	If you are following advice given earlier, you will be designing experiments or planning how to make careful observations. Although you may use some standard protocols, there should ideally be a personal input to the experimental design, even if you are working in a research laboratory. You must show that you understand the theory behind any methods that you use, and the limitations and uncertainties involved – if you do not then the methods are too complex!
Collect the data that you need	Remember the things that you have been taught when planning experiments for Internal Assessment – variables must be controlled and repeats are needed to allow you to assess the reliability of your data. If you aren't doing your own experiments, you must obtain the published results of experiments, not just the conclusions that were drawn from these experimental results.

Writing up your essay

Write an introduction	This can be quite brief. There is no need to include large amounts of background material, especially if it is straightforward biology. Instead, say why the topic is worth writing about and give the background information that is needed to understand the essay. The introduction should make it clear how and why you have chosen your research question. You should, of course, state your research question precisely, either in italics or bold type.
Describe your methods	This section shouldn't be very long – if it is then your methods were probably too complex. Explain clearly and fully what you did and why. You should include enough detail to allow your experiments to be repeated. Make clear how the experiments tested your hypothesis or gave the evidence needed to answer the research question. Explain the limitations and uncertainties that were caused by the methods that you used.
Display your results	Use clear tables or other formats to display the data that you have obtained – the results of your experiments. You only need to put raw data into an appendix if you have huge amounts of it. Use graphs or other charts to display the most significant features of the data, for example mean results. If you are using data from other scientists, you should display and manipulate it in an original way.
Analyse your data	This should be a long and detailed section of the essay. You should discuss whether the data is reliable – were the repeats close? do the results show a consistent trend? What confidence level do statistical tests show? Were there errors or uncertainties that had an impact on your results? Then use your understanding of the topic to discuss possible explanations for any trends, with reasons for rejecting or accepting them.
Draw your conclusions	Only a short section is needed here. It should not include new information or views different from those expressed in earlier sections of the essay. Instead, you should sum up what answer you have found to your research question or whether your hypothesis is supported or undermined. Your conclusions should be based only on the data that you have obtained and analysed. You should make it clear what the unresolved issues are, and suggest how they could be investigated.
Write an abstract	You must summarize your whole essay in 300 words. You must include your research question clearly, how you investigated it and what conclusion you reached. The usual purpose of an abstract is to give the reader a quick impression of the contents of a long article so that he or she can decide whether it is worth reading or not! Obviously your essay will be well worth reading!
Add the finishing touches	You now need to write a contents page and a bibliography. The contents page lists the sections of the essay with the number of the page on which each section begins. The bibliography is a numbered list of the published sources that you used. You should put a reference in the text of your essay, in the form of a superscript number, wherever you have used information from these sources. Proof-read your essay to check for spelling or grammatical mistakes.

If you want to do well in your final exams, you must prepare for them very carefully in the weeks beforehand.

The most important task is to memorize all the facts that you have been taught. For a high grade, you will need a comprehensive knowledge of them. You will need to spend many hours on revision and find tactics that work for you.

You should also practise answering exam questions. You can use the questions at the end of topics in this book, after you have revised each topic. Your teacher should also give you some past exam papers to try.

There are three styles of question in IB Biology exams.
- **Multiple choice questions** – These are questions where you choose one of four possible answers. Read all of them before choosing one. If you cannot decide on one answer, try to eliminate answers that are obviously wrong to narrow down the possibilities. Leave difficult questions until you have answered the straightforward ones. Give an answer to every question – marks are not deducted for wrong answers.
- **Structured questions** – These questions are broken up into small sections, each of which you answer in the space or on the lines provided. If you run out of space, continue your answer elsewhere on the paper – it will be marked as long as you indicate clearly what you have done. The number of marks for each section is indicated and this helps show you how detailed your answer needs to be. Some structured questions involve data analysis. Look through the data questions in this book to see some of the ways in which data can be presented. You should always study the data very carefully before answering the questions, for example the scales and labelling on the axes of graphs. If there are calculations, remember to show your working and give units with your answer, for example grams or millimetres.
- **Free response questions** – These questions require long and detailed answers on lined paper. You can decide what style of answer to give. Usually continuous prose is best, but sometimes ideas can be shown on a table or on a carefully annotated diagram. There may be a choice of free response question. Read the whole of each question before making your choice.
  There will be marks for the quality of construction of your answer. If the question is divided up into sections (a), (b) and so on, you must answer it in these sections. Try to express your ideas clearly so that the examiner understands what you mean. Plan out your answer on scrap paper, so that you arrange your ideas in a logical sequence and do not include irrelevant material.
  As with all questions, you must write legibly or the examiner may not be able to mark your work. This may mean that you have to write more slowly than normal.

If you do revise carefully and build up a comprehensive knowledge of the facts on the syllabus you should find many of the questions straightforward. This is because, in IB Biology exams, 50% of the marks are for simple factual recall. These questions will start with words like *list, state, outline* or *describe*. The other 50% of the marks involve more than simple factual recall – they involve expressing ideas that are more complex or involve using your knowledge to develop an answer that you probably haven't been taught.

The word at the start of each question tells you what to do. These words are therefore called command terms.

Explain – Sometimes this involves giving the mechanism behind something – often a logical chain of events, each one causing the next. This is a 'how' sort of explanation. A key word is often 'therefore'. Sometimes it involves giving the reasons or causes for something. This is a 'why' sort of explanation. A key word is often 'because'.

Discuss – There won't be a simple straightforward answer to these questions. Sometimes your answer should include arguments for and against something. Try to give a balanced account. Sometimes your answer should consist of a series of alternative hypotheses – you could indicate how likely each one is but you don't need to make a final choice.

Suggest – Don't expect to have been taught the answer to these questions. Use your overall biological understanding to find answers – as long as they are possible, they will get a mark.

Compare – This type of question involves assessing how similar or different two or more things are. You cannot do this by describing the things separately. Every point that you make should be a similarity or a difference. There may be more similarities or more differences – all of them are relevant.

Distinguish – This is similar to a compare question, except that only differences need to be included in your answer. The key word in this type of question is often 'whereas'.

In both compare and distinguish questions a table is often the best way to arrange your answer. Use the columns of the table for the things that you are comparing and the rows for the individual similarities or differences.

Evaluate – This usually involves assessing the value, importance or effects of something. You might have to assess how useful a technique is, or how useful a model is in helping to explain something. You might have to assess the expected impacts of something on the environment. Whatever it is that you are evaluating, you will probably have to use your judgement in composing your answer.

There are other command terms that are used in questions, but they are more straightforward and you are unlikely to have difficulty in understanding what sort of answer to give.

TOPICS 1 AND 2 STATISTICAL ANALYSIS AND CELLS

1 (a) X = rough endoplasmic reticulum Y = mitochondria (b) nuclear membrane is the curved structure on the left-hand side
(c) proteins because there is rough endoplasmic reticulum with ribosomes which make protein; ATP because there are many mitochondria which make ATP.

2 (a)(i) phospholipid (ii) head is hydrophilic and tails are hydrophobic (b)(i) II is integral (ii) any two of: III is a pump protein; transfers specific substances; uses energy from ATP to move substance against the concentration gradient.

3 (a)(i) 12.4; %; (ii) 5.22 (b)(i) positive; correlation; (ii) does not prove obesity causes high blood pressure; correlation does not establish a causal relationship; high blood pressure may be caused by something else that also caused obesity;

TOPIC 3 THE CHEMISTRY OF LIFE

1 (a)(i) DNA (ii) DNA (iii) RNA (b) experimental error (c)(i) DNA is double stranded; A pairs with T and C pairs with G; one base in each pair is therefore A or G, so A + G = 50%; (ii) any two of A = T; C = G; C + G = 50%; A + G / C + G = 1.00
(d)(i) influenza virus (ii) RNA contains uracil instead of thymine; single stranded so amounts of G and C not equal.

2 (a)(i) CO_2 concentration falls in the light and rises in the dark; (ii) CO_2 concentration falls when it is warmer and rises when it is cooler; (b) CO_2 concentration is more closely related to light intensity; when there is a temporary dark period during the third day but it stays warm pH drops so CO_2 concentration rises; (c)(i) respiration producing CO_2; (ii) photosynthesis causing CO_2 uptake;

3 (a) radical/variable portion of the amino acid (b) C – N bond; O= linked to C and H- linked to N (c) 70S ribosomes in prokaryotes vs. 80S ribosomes in eukaryotes; free ribosomes in prokaryotes vs. ribosomes sometimes linked to rough endoplasmic reticulum in eukaryotes.

TOPIC 4 GENETICS

1 (a) O group individual must be genotype ii because it is due to a recessive allele; B group individual in generation 2 must be I^Bi because the parent that was blood group A could not have passed on I^B; B group individual in generation 3 must have been I^Bi because the O group parent must have passed on i;
(b) parents could have been group O; parents could have been group A with genotype I^Ai; parents could have been group B with genotype I^Bi (c) blood transfusion.

2 (a) $C^r C^r$ $C^W C^W$ and $C^r C^W$ (b) The allele for red flowers is dominant in peas but codominant in *Mirabilis* (c) gametes C^r and C^W; genotypes $C^r C^r$ $C^r C^W$ $C^W C^r$ and $C^W C^W$; phenotypes red pink pink and white, respectively.

3 (a) a group of organisms with identical genotypes (b) nucleus removed from a cell in an adult organism; nucleus removed from an egg cell and replaced with the nucleus from the adult animal (c)(i) fragments had moved down; larger fragments are nearer the top and move more slowly; (ii) culture cells have the same profile as udder cells as they have the same pattern of bands; Dolly's blood cells have the same profile as the udder/culture cells as they have the same pattern of bands; Dolly was cloned from the udder cells; sheep 1–12 are genetically different;

TOPIC 5 ECOLOGY AND EVOLUTION

1 (a) sigmoid/S-shaped (b)(i) line reaching a plateau by year 8 (ii) any two of: food supply; predation; breeding sites; disease (c)(i) population would have reached carrying capacity more quickly (ii) carrying capacity would have been the same.

2 (a) I = secondary consumers II = primary consumers III = producers (b) chemical energy (c) arrow from the sun to box III
(d) any two of: arrows represent energy losses; heat produced because energy transformations are never 100% efficient; energy not passed along the food chain to another organism.

3 (a) methane causes an increase in the Earth's temperature by the greenhouse effect; temperature only increases as a result of an increase in atmospheric methane; methane emissions to the atmosphere must be greater than losses (b) methane emission is a natural process, e.g. swamps and marshes; humans cause methane emission, e.g. coal burning/cattle and sheep/rice paddies; most emissions are caused by humans/humans have increased emissions considerably (c) any three of: drain swamps and marshes; reduce cattle and sheep farming; stop growing rice in paddies; control releases of natural gas; reduce burning of coal; prevent forest fires/burning of biomass.

TOPIC 6 HUMAN HEALTH AND PHYSIOLOGY

1 (a) T_b is higher than T_a; T_b is constant whereas T_a is decreasing (b) heat absorbed from the environment; heat generated by cell respiration (c) active during darkness because it maintains constant high body temperature as a result of cell respiration.

2 (a)(i) ingestion of pathogens (ii) in blood; in body tissues (b) to allow the production of many different types of antibody; to fight many different diseases.

3 (a) I = trachea II = bronchioles (b) maintains concentration gradients of oxygen and CO_2 between air in alveoli and blood; ensures rapid diffusion/gaseous exchange (c) alveolus wall consisting of very thin cells; blood capillaries adjacent to alveolus; bronchiole connected to alveolus; diameter of alveolus indicated;

TOPIC 7 NUCLEIC ACIDS AND PROTEINS

1 (a)(i) higher than 40 °C; initial rate was faster; then reaction stopped due to denaturation (ii) lower temperature than 40 °C because the rate is slower; 30 °C because the rate is half that at 40 °C (b)(i) curve drawn above the curve W (similar to curve Y) (ii) curve drawn above the curve W; gradient of curve decreasing markedly with time showing increasing inhibition as the substrate concentration falls.

2 (a) 3′ terminal is deoxyribose/ribose to which a nucleotide can be linked; 5′ terminal is phosphate group to which a nucleotide can be linked (b) Any 3 of : purines and pyrimidines are both bases; both are part of nucleotides; A and G are purines and C and T are pyrimidines; purines can only pair with pyrimidines in DNA; purines have two rings and purines only one ring
(c) Any 5 of: DNA is transcribed; mRNA is translated; RNA is produced by transcription; polypeptides are produced by translation; transcription is done by RNA polymerase; translation is done by ribosomes;

3 (a) globular (b) number and sequence of amino acids (c)(i) X is a beta pleated sheet and Y is alpha helix (ii) hydrogen bonding (d) any two of: tertiary structure determines the enzyme's shape; determines the active site's shape; makes the enzyme substrate specific; shape ensures that when the substrate binds it is distorted/induced fit.

TOPIC 8 CELL RESPIRATION AND PHOTOSYNTHESIS

1 (a) Any two of: double membrane; cristae/infoldings of inner membrane; ovoid shape; (b) double outer membrane shown; inner membrane shown folded in to form a crista (c)(i) label indicating the matrix (ii) label indicating the inner membrane/cristae (iii) label indicating the cytoplasm outside the mitochondria.

2 (a) peaks in the red and blue sections of the spectrum; minimum in the green section at about half of maximal rate (b) action and absorption spectra are closely correlated; because pigments absorb the light energy used in photosynthesis; the more light absorbed at a wavelength the more photosynthesis.

3 (a) oxidative phosphorylation and photophosphorylation (b) barrier to proton movement; allows a proton gradient to develop; location of ATP synthase; (c) plasma membrane;

TOPIC 9 PLANT SCIENCE

1 (a)(i) more Pfr (ii) more Pfr (b) Pfr slowly reverts to Pr in darkness; timing is based on the amount of conversion; (c) Pfr promotes flowering in long-day plants; and inhibits it in short-day plants;

2 (a) Any three of: parallel versus net-like veins; vascular bundles distributed through stem versus in a ring; one versus two cotyledons; floral organs in 3s versus in 4s or 5s; unbranched versus branched roots; (b) apical meristem increases length of the stem; lateral meristem increases width of stem; (c) monocots cannot thicken their stems; cannot grow into large trees; less opportunity for branching;

3 (a)(i) 6 pm to 6 am/sunset to sunrise (ii) 6 am to 4.30 pm (b) CAM plant is the xerophyte because it opens its stomata at night; less water loss during cooler conditions in the night (c)(i) partial closure between 11 am and 12 am; followed by reopening (ii) plant needs to limit transpiration during the hottest part of the day.

TOPIC 10 GENETICS

1 (a) polygenic (b) AaBb; blue-flowered (c) all gametes shown with one allele of each gene only; four homozygous genotypes shown AABB AAbb aaBB and aabb; four double heterozygous genotypes shown AaBb; eight other genotypes shown AABb AAbB aaBb aabB AaBB aABB Aabb and aAbb; all sixteen phenotypes indicated (d) 9 blue 3 red and 4 white (e) gene A converts white to red and gene B converts red to blue.

2 reassortment of genes into different combinations from those of the parents (b) black body long wing; grey body vestigial wing (c) genes are linked/found on the same chromosome; parental combinations are kept together; unless there is a cross-over between the genes (d) any two of: find which chromosome a gene is located on; identify all of the genes in a linkage group; estimate how far apart the loci of genes on a chromosome are.

3 (a) first; prophase; (b)(i) four chromatids (ii) five chiasmata; (c) breakage of chromatids; rejoining of non-sister chromatids; exchange of material between chromatids;

TOPIC 11 HUMAN HEALTH AND PHYSIOLOGY

1 (a) excretion is removal from the body of waste products; waste products of metabolism (b) Any two of: protein in blood plasma but not urine; glucose in blood but not in urine; higher concentration of urea/waste products of metabolism in urine than blood plasma; composition of urine is more variable than blood plasma (c) loop of Henle makes medulla hypotonic by raising sodium/mineral ion concentration; allows production of hypertonic urine (d) basement membrane of glomerulus/filtration slits;

2 (a)(i) actin (ii) regions II and III (b)(i) II would increase in length. (ii) I and III would increase in length.

3 (a) Any four of: both contain a haploid nucleus; both have a plasma membrane; the sperm has a tail but the egg does not; the egg has much more cytoplasm; mitochondria in sperm are helical but in eggs are ovoid; the egg has cortical granules but the sperm does not; (b) stimulates gametogenesis in both men and women; promotes development of follicles in women and primary spermatocytes in men; stimulates estrogen secretion in women but not testosterone in men; (c) both stimulate the development of the corpus luteum; both stimulate the secretion of progesterone; before fertilization by LH and after fertilization by HCG;

OPTION A – HUMAN NUTRITION AND HEALTH

1 (a)(i) 24.3 (ii) 22.6 (b)(i) 24.2 and 22.5 (ii) values are very close (c) below 18.5 is underweight; 25–30 is overweight; above 30 is obese;

2 (a)(i) monounsaturated fatty acids have one double bond and polyunsaturated have more than one; (ii) trans fatty acids have hydrogen bonded to different sides of double bonded carbon atoms versus cis fatty acids have hydrogen bonded to the same side; (b) saturated fatty acids are linked with increased blood cholesterol; cholesterol/saturated fatty acids are linked with atheroma; correlation between saturated fat intake and CHD;

3 (a) one mark for any two of: protein; fats; carbohydrate; minerals; vitamins; water; (b) Any three of: human milk contains the ideal combination of nutrients for human babies; breast-feeding helps with bonding; breast milk contains antibodies; human milk does not cause allergies (c) milk production from cattle involves separating calves from their mothers when they are very young; also involves slaughter of calves/young cattle;

OPTION B – PHYSIOLOGY OF EXERCISE

1 (a) humans store less oxygen per kg of body tissue (b) Any three of: both store most in blood; seal stores a higher proportion in blood than human; smallest proportion stored in lung of seal but muscle of human; human stores higher proportion in lung than seal; seal stores higher proportion in muscle than human (c) any three of: size of muscle; ratio of fast and slow fibres; concentration of myoglobin in muscle; amount of blood in muscle.

2 (a) thin actin and thick myosin filaments shown; actin filaments attached to Z discs; actin filaments interdigitating with myosin; (b) ATP provides energy; for myosin heads; to detach from actin and recock (c) cardiac output is higher when muscles are contracting;

3 (a) two; (b) discs bulge; soft pulpy core of disc is protruding; (c) (white matter of) spinal cord; (d) abnormal neck movements; heavy loads;

OPTION C – CELLS AND ENERGY

1 (a) CO_2 concentration (b)(i) temperature; rate of photosynthesis rises as temperature rises; (ii) temperature controls the rate of photosynthesis between 35 and 40 °C; but is not the factor nearest to its minimum level/is supra-optimal (c) light is the limiting factor at low light intensity; temperature therefore does not affect the rate of photosynthesis.

2 (a) (i) malonate inhibits succinate dehydrogenase/other example (ii) copper/mercury/silver ions/other example; (b) similarity: both types of inhibitor reduce the rate of catalysis; difference between competitive and non-competitive: inhibitor structure similar to substrate vs. not similar/inhibitor binds to active site vs. binds elsewhere.

3 (a) pyruvate; acetyl group; (b) both are CO_2; (c) NADH + H$^+$ (d) Krebs cycle;

OPTION D – EVOLUTION

1 (a)(i) positive correlation (ii) any two of: primate brains are larger in relation to body mass; but there is much variation; largest primates have relatively small brains (iii) any two of: scattergram shows that human brain has the largest size; primates with a larger body mass have a smaller brain; human brain mass is furthest above the line of best fit (b) easier to climb trees/speed of movement/less food needed.

2 (a)(i) 2 (ii) 6 (iii) 9 (iv) 6 (v) 10 (vi) 7 (b) cladogram with four species; first split between rabbit and other three species; second split between lemur and other two species; final split between humans and orang utans;

3 (a) $p^2 + 2pq + q^2 = 1$ and q^2 is the frequency of homozygous recessives; frequency = 0.23/23% (b) 35% (c) carriers have increased resistance to malaria; selective advantage over homozygous dominants so the sickle cell allele survives.

OPTION E – NEUROBIOLOGY AND BEHAVIOUR

1 (a) receptor protein; each receptor protein's shape is complementary to a specific odorant; (b) Any three of: G protein activates the enzyme adenylyl cyclase; enzyme converts ATP to CAMP; CAMP causes calcium channel to open; calcium causes chloride channel to open (c) membrane of chemoreceptor cell depolarizes/action potential created/chemoreceptor cell passes an impulse to a sensory neuron.

2 (a) photoreceptors (b) in sensory neurons from the retina to the brain; in motor neurons from the brain to the circular muscle fibres in the iris (c) no response when a light is shone into eye of unconscious patient indicates damage to the brainstem.

3 (a) supporting the hair cells/reticular lamina; (b)(i) three rows; small medium and longer stereocilia; arranged in a W shape; (ii) 1.2 µm; (iii) longer perceive lower frequency sounds; (c)(i) amplifies sounds; for the inner hair cells to perceive more easily; (ii) in the plasma membrane; mitochondria close to the edge of the cell;

OPTION F – MICROBES AND BIOTECHNOLOGY

1 (a) bacteria numbers increase; bacteria use oxygen in aerobic cell respiration; (b) bacteria decompose raw sewage; ammonia and phosphate are released during decomposition; (c) ammonia is converted to nitrate; by nitrifying bacteria; (d) eutrophication; nutrients stimulate growth of photosynthetic bacteria/algae; (e) bacteria consumed by other organisms; raw sewage has all been decomposed;

2 (a) synthesis of DNA/cDNA; from RNA (b)(i) retroviruses (ii) HIV (c) any three of: mRNA can be obtained quite easily; genes can be hard to find; gene consisting of DNA can be made from RNA; no introns in the gene using reverse transcriptase; gene can be inserted into other organisms; cDNA/probes can be used to locate other genes;

3 (a)(i) protein coat/capsid (ii) nucleic acid/genes; (b) insert gene into viral DNA; virus acts as a vector for the gene; (c) intracellular; all viruses rely on a host cell for most of their processes;

OPTION G – ECOLOGY AND CONSERVATION

1 (a)(i) *S. marcescens* feeds on the nutrients so more grow at high nutrient levels (ii) *C. striatum* reduce the numbers by predation (iii) *D. nasutum* increases the numbers because it feeds on *C. striatum*; which reduces the predation of *S. marcescens* (b) low population of *S. marcescens* at low nutrient levels; therefore very low levels of *C. striatum* on which *D. nastum* feeds (c) longer food chain with higher nutrient levels.

2 (a) indicator species need particular environmental conditions; can be used to give a measure of pollution levels/levels of an environmental variable (b) any two of: captive breeding and release of endangered species; growth of endangered plants in botanic gardens; storage of frozen seeds of endangered species in seed banks.

3 (a) mercury inputs reduced; (b) increases/decreases when river flow rates rise/fall; high rainfall leaches more mercury out; (c) death of organisms in higher trophic levels; toxic effects for humans consuming fish from higher trophic levels;

OPTION H – FURTHER HUMAN PHYSIOLOGY

1 (a) *H. pylori* is implicated as a cause of stomach ulcers/cancer; antibodies show that the patient has been infected with *H. pylori*; (b) Any two of: incidence of stomach ulcers and cancer is higher in patients who had been infected with *H. pylori*; all patients with stomach cancer had been infected with *H. pylori*; some patients with stomach ulcers had not been infected so there must be alternative causes; correlation does not prove causation; (c)(i) higher incidence of duodenal inflammation in patients who had been infected; higher incidence of oesophagus inflammation in patients who had not been infected; (ii) *H. pylori* infects the stomach; toxins produced by *H. pylori* will pass on to the duodenum, not back to the oesophagus;

2 (a) SAN/pacemaker sends out a signal; signal spreads out through the walls of the atria (b) any two of: lub dup sounds made when valves close; closing valve causes vibration of blood in ventricle; rushing sounds due to flow of blood.

3 (a) V_E increases as carbon dioxide concentration increases; greater increases in V_E with successive increases in carbon dioxide concentration; (b) increases in carbon dioxide concentration in inspired air increase the blood concentration; detected by chemosensors in aorta/carotid artery; impulses sent to ventilation centre of brain/medulla; (c) further increases in V_E; until maximal V_E is reached;

INDEX TO ASSESSMENT STATEMENTS

Index

A

abiotic factors 152
ABO blood groups 27
abscisic acid 83
absorption of food 165
absorption spectrum 77
acetyl coenzyme A 74
acrosome reaction 106
actin 100
action potential 53
action spectrum 77
activation energy 69
active immunity 97
active management 158
active sites 18, 69, 70
active transport 9, 84
adaptation 38
adaptive radiation 125
addiction 137
adenine 60
adenosine triphosphate 20
ADH 102, 162
adrenalin 48
aerobic cell respiration 20, 75, 117
AIDS 50
albinism 29
alcohol 137
alcohol abuse 166
algal blooms 144, 145
alien species 157
allele frequencies 123, 128
alleles 23, 26
allopatric speciation 124
allostery 71
alpha helix 66
altitude 169
altruism 140
alveoli 51
amino acids 14, 62, 66, 67
amino acids in food 11
amniocentesis 25, 107
Amoeba 143
amphetamines 137
amylase 147
Anabaena 148
anabolic steroids 116
anaerobic cell respiration 20, 74, 117
analogous structures 129
angiospermophytes 35
animal cells 7
animal experiments 138
annelida 35
anorexia nervosa 113
anterior pituitary 162
anthrax 98
antibiotic resistance 39
antibiotics 49, 149
antibodies 49, 50, 96, 97
anticodon 17, 63
anti-diuretic hormone 102, 162
antigens 49, 96
antisense strand 62
antiseptics 149
apical meristems 87
appetite control 112
archaea 142, 145
Ardipithecus 126, 127

arteries 48
arterioles 55
arthropoda 35
Asian flu 150
Aspergillus 147
assimilation 47
asthma   170
atherosclerosis 168
atmosphere 123
ATP 20, 73, 78
ATP synthase 75
ATP synthesis 74, 75
ATPase 75
Australopithecus 126, 127
autonomic nervous system 137
autosomal linkage 92
autotrophs 41
auxin 87
AV (atrioventricular) node 167

B

bacteria 6, 49, 142
bacterial infection 149
base pairing 16, 60, 62
base substitution 23
B-cells 96
beer production 147
bees 140
behavioural isolation 124
benzodiazepines 137
beta pleated sheets 66
bile 166
bile salts 164
binomial system 34
biochemical oxygen demand 144
biodiversity 156
biofilms 143
biogas 145
biogeographical features 158
biological control 157
biomagnification 156
biomass 21, 154
biomes 155
bioreactors 145
bioremediation 148
biosphere 155
biotic factors 152
biotic index 158
bipedalism 126
birdsong 135
birth 107
bladder 56
blood 48
blood clotting 98
blood groups 27
BOD 144
body mass index 112
body temperature 55
Bohr shift 169
bones 99
botanic gardens 158
bovine spongiform encephalopathy 150
brain size 127
bread making 147
breast-feeding 113
bryophytes 35

BSE 150
Bt maize 31
bulbs 86
buoyancy 15

C

calcium 14, 52
Calvin cycle 79
CAM plants 83
cambium 87
cancer 11
cannabis 137
capillaries 48, 55
capsid 143
captive breeding 158
capture–mark–release method 159
carbaminohemoglobin 170
carbohydrates 15
carbohydrates in food 112
carbon cycle 43
carbon dioxide transport 170
carbon fixation 21, 43, 79
carbonic anhydrase 170
cardiac cycle 167
cardiac output 119
cardiovascular systems 119
Carolina parakeet 160
carriers 28
carrying capacity 36
cartilage 99
cDNA 146
cell division 11, 20, 73, 74
cell theory 3
cell walls 6, 7, 142
cells 3, 10, 15, 114
central nervous system 52, 136
centrioles 11
centromere 23
cerebellum 138
cerebral hemispheres 138
CFCs 157
channels 8
CHD 111, 114, 168
chemiosmosis 75, 78
chemoautotrophs 148
chemoheterotrophs 148
chemosensors 170
chiasmata 93
childbirth 107
Chlamydia 149
Chlorella 143
chloride shift 170
chlorophyll 21, 77, 78
chloroplasts 7, 79, 80, 123
cholesterol 114, 166, 168
chorion 108
chorionic villus sampling 25
chromatids 23
chromosomes 23
cirrhosis 166
CJD 150
clade 130
cladistics 130
cladograms 130
class 35
classification 3, 129, 130, 142

climograph 155
clinical obesity 112
clonal selection 97
clone 32
cloning 32
clotting 98
cnidaria 35
CNS 52, 136
cocaine 137
cochlea 133
codominance 27, 90
co-dominant alleles 27, 28
codons 17, 62
collagen 68
colour blindness 28
combustion 43
communities 40
competition 153
competitive exclusion 153
competitive inhibition 70
condensation reactions 15
conditioning 135
cone cells 134
coniferophytes 35
conifers 35
conjugated proteins 67
conservation 156, 158
consumers 41, 42
continuous variation 91
convergent evolution 125
copulation 107
cornea 134
coronary heart disease 111, 114, 168
coronary thrombosis 168
corpus luteum 57, 104
correlation 2
cortical reaction 106
creatine phosphate 117
creation of life 122
Creutzfeld-Jacob disease 150
cristae 76
crossing over 92, 93
cultural evolution 126
cyanobacteria 148
cyclic photophosphorylation 81
cystic fibrosis 128
cytokinesis 11
cytoplasm 3, 6, 7
cytosine 60

D
Darwin 38
decarboxylation 74
decision making 136
deficiency diseases 110
denaturation 18
denitrification 144, 145
deoxyribonucleic acid 16
depolarization 53
desert 155
detoxification 166
detrivores 41
diabetes 55, 113
diaphragm 51
dicotyledons 86
dietary fibre 114
dietary supplementation 110
diets 110
differentiation 3, 4
diffusion 9
digestion 47, 163

digestion 163
digestive enzymes 163, 164
digestive juices 163
dihybrid crosses 89, 90
diploid 24
disaccharides 15
disease 49, 50
disinfectants 149
dislocation of joints 119
distribution of animals 152
distribution of plants 152
disulfide bridges 66
diversity index 156
DNA 16, 60
DNA fingerprinting 30
DNA polymerase 60
DNA profiling 30
DNA replication 16
Dolly the sheep 32
domains 142
dominant alleles 16
dopamine 137
Down syndrome 25
drugs 137
drugs in sport 116

E
ear 133
ecological efficiency 42
ecological niches 153
ecological succession 155
ecology 43
ecosystems 43
effectors 132
egestion 47
egg cell 104, 105
eggs in the diet 114
elbow 99
electron transport chain 75
elements 14
embryos 58
emergent properties 3
emulsification of fats 164
end product inhibition 71
endocrine system 54
endocytosis 10, 165
endoplasmic reticulum 7, 10
endorphins 139
endosymbiotic theory 123
endotoxins 149
energy 20
energy efficiency 42
energy flow 41
energy in food 112
energy losses 41, 42
energy pyramids 42, 154
energy requirements 112
energy storage 15
environmental monitoring 158
enzyme inhibition 70, 71
enzyme specificity 18
enzymes 18, 19, 69
epidemics 150
epistasis 90
epithelium 165
EPO 119
ER 7
error bars 1
erythrocytes 49
erythropoietin 119
estimating fish stocks 160

estrogen 57, 108
ethanol 20
eubacteria 142, 143
Euglena 143
eukaryotes 7, 61, 142
eutrophication 144, 145
evolution 37, 38, 39, 125, 127
evolutionary clocks 129
ex situ conservation 158
exagerrated traits 140
excitatory drugs 137
excitatory synapses 136
excretion 101, 102
exercise and ventilation 170
exergonic reactions 69
exocrine glands 163
exocytosis 10
exons 61
exotoxins 149
extracellular infections 149
extracellular material 3, 10
eye 134

F
F_1 hybrids 16
facilitated diffusion 9
family 35
fast and slow muscle 116
fatty acids 14, 111
ferns 35
fertilization 58, 85, 106, 107
fetal development 107
fetal hemoglobin 169
fibre 114
fibrinogen 98
fibrous proteins 68
filicinophytes 35
fish conservation 160
fish stocks 160
fish yields 160
fitness 116
flagella 6, 143
flatworms 35
flowering 87
flowers 85, 86
fluid mosaic model 8
fMRI 138
follicle 56, 57, 104
food chains 40
food miles 114
food poisoning 147
food preservation 147
food webs 42
foraging behaviour 139
forensics 30
fossil fuels 43
fossilization 126, 127
fossils 37
fructose 15
FSH 56, 57, 58
functional magnetic resonance imaging 138
fundamental niches 153

G
gametogenesis 103, 104, 105
gas exchange 51, 169
gastric juice 164
gel electrophoresis 30
gender 28
gene interaction 90

gene linkage 92, 93
gene mutation 23
gene pools 123, 128
gene therapy 146
genes 18, 23
genetic code 17, 62
genetic diseases 23, 29
genetic evolution 127
genetic modification 31
genetic variation 24
genome 23
genotype 16
genus 34
geographical isolation 124
germination 85
germ-line therapy 146
giberellin 85
glands 163
global warming 44
globular proteins 68
glomerulus 101
glucagon 55
glucose 14, 15, 55
glycerides 15
glycerol 14
glycogen 7, 15, 117
glycolysis 73
glycoproteins 8, 10
GMO 31
goitre 110
Golgi apparatus 7, 10
gradualism 125
Gram stain 142
grassland 155
greenhouse effect 44
gross production 154
guanine 60
guard cells 83

H
Haber process 144
habitat 45
haemoglobin 67, 68, 166, 169, 170
half-life 125
halophiles 142
haploid 24
Hardy–Weinberg equation 128
Hardy–Weinberg principle 128
HCG 58, 108
hearing 133
heart 48, 119, 167
heart action 167
heart attacks 168
heart beat 167
heart rate 119
helicase 60
Helicobacter pylori 164
helper T-cells 96
hemoglobin 67, 68, 166, 169, 170
hemophilia 28
hepatic blood vessels 166
hepatocytes 166
herbivory 153
heterotrophs 41
heterozygous 16
hip joint 99
histone 61
HIV 50, 146
homeostasis 54, 55
hominids 126, 127
Homo 126, 127

homologous 24
homologous structures 129
homozygous 16
honey 114
honey bees 140
hormones 8, 162
human ancestors 126, 127
human classification 127, 129
human diets 110
human evolution 126, 127
human genome project 32
human impacts 45
human milk 113
human origins 126, 127
human reproduction 103
hybridoma cells 98
hydrogen bonds 13, 16, 60
hydrogen carbonate 170
hydrolysis reactions 15
hypothalamus 55, 112, 138, 162

I
IDD (iodine deficiency disorder) 110
identification 34
ileum 165
immunization 97
immunity 97
immunoglobins 68
in situ conservation 158
in vitro fertilization 58
independent assortment 89, 92
indicator species 158
induced fit hypothesis 69
infections 49, 149
infertility 58
influenza 149
inhibitors 70, 71
inhibitory drugs 137
inhibitory synapses 136
initiation of translation 65
injuries in sport 119
innate behaviour 135
inoculation 97
inorganic compounds 14
insect pollination 85
instinct 135
insulin 31, 55
international conservation 160
interphase 11
interspecific competition 157
intervertebral discs 119, 120
intestines 47
intracellular infections 149
intramolecular bonding 66
introns 61
iodine 110
iris 134, 139
iron 14
IVF 58

J
joints 99

K
karyotypes 24, 25
keys 34
kidney structure 101
kinesis 135
Krebs cycle 74
K-strategies 159

L
lactase 19
lactate 20, 117
lactic acid 20, 117
lactose 15, 19
lateral meristems 87
LDL 168
learned behaviour 135
leaves 83, 86
leptin 112
lesions 138
leukocytes 49
LH 56, 57
ligaments 99, 119
ligase 60
light-dependent reactions 77, 78
light-independent reactions 77, 79
limiting factors 81
link reaction 74
linkage 92
lipase 68
lipid digestion 164
lipids 15
lipids in food 112, 114
liver 55, 166
liver cell 7
liver damage 166
loop of Henle 102
lung capacity 118
lungs 51
lymphocyte 49, 50
lysosomes 7
lytic life cycles 149

M
macroevolution 126
macrophages 96
magnification 5
malaria 98, 150
malnutrition 111
maltose 85
management of wildlife reserves 158
mean 1
meat in the diet 114
medulla oblongata 138
meiosis 24, 92, 93, 94
melanisim 39
membrane proteins 8, 68
membranes 8
Mendel 16, 89
menstrual cycle 57
menstruation 57
meristems 87
metabolic pathways 71
metabolism 71
methane generation 145
methanogens 142
microevolution 127
micrometres 5
microvilli 47, 102, 108, 165
migration 132
milk 113, 114
Miller and Urey 122
mineral absorption 84
mineral elements 14
mineral ion uptake 84
minerals in food 110
mitochondria 7, 74, 75, 76, 123
mitosis 11
mollusca 35
monitoring environments 158

184